# Practical
# Math

**GRADE 5**

*Advanced Version*

Kwang S. Ko, Ph.D.

1 *A SELF-STUDY GUIDE*
2 *EXERCISES*
3 *SELF-TESTS*
4 *SOLVING PROBLEMS*
5 *FULL ANSWER KEY*

Request for information should be addressed to 7722 Camino Noguera, San Diego, CA 92122
Visit our website at www.iqmaths.com


ISBN: 978-1523363018


Printed in the United States of America

10 9 8 7 6 5 4 3 2

# CONTENTS

# CHAPTER 1
# Number Sense

In this chapter, you will learn number sense to compute addition, subtraction, multiplication, and division expressed in whole numbers.

## 1. Place Value

**1-1.** Understanding Place Value

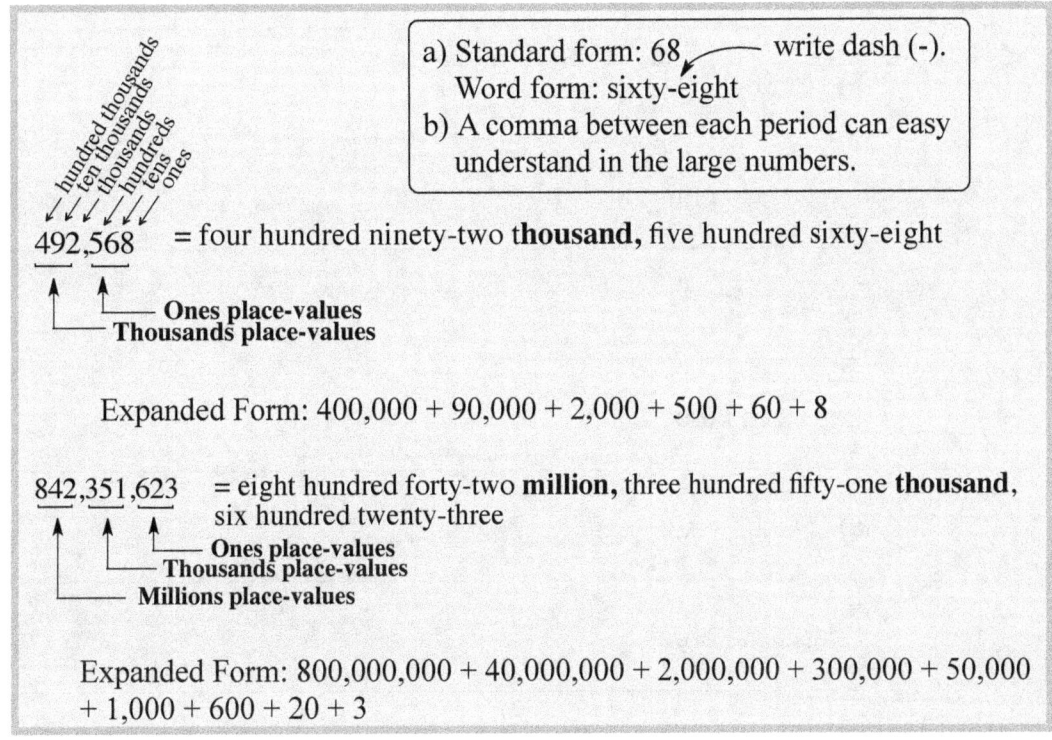

Expanded Form: $400,000 + 90,000 + 2,000 + 500 + 60 + 8$

842,351,623 = eight hundred forty-two **million,** three hundred fifty-one **thousand,** six hundred twenty-three

Ones place-values
Thousands place-values
Millions place-values

Expanded Form: $800,000,000 + 40,000,000 + 2,000,000 + 300,000 + 50,000 + 1,000 + 600 + 20 + 3$

**1-2.** Place Value System

| Standard form | Word form | Factors | Exponent form |
|---|---|---|---|
| 1,000 | one **thousand** | 10 x 10 x 10 | $10^3$ |
| 10,000 | ten **thousand** | 10 x 10 x 10 x 10 | $10^4$ |
| 100,000 | one hundred **thousand** | 10 x 10 x 10 x 10 x 10 | $10^5$ |
| 1,000,000 | one **million** | 10 x 10 x 10 x 10 x 10 x 10 | $10^6$ |
| 10,000,000 | ten **million** | 10 x 10 x 10 x 10 x 10 x 10 x 10 | $10^7$ |
| 100,000,000 | one hundred **million** | 10 x 10 x 10 x 10 x 10 x 10 x 10 x 10 | $10^8$ |
| 1,000,000,000 | one **billion** | 10 x 10 x 10 x 10 x 10 x 10 x 10 x 10 x 10 | $10^9$ |

**Exercises 1**   Write each number in its word and expanded forms.

**1)** 253

_____

Expanded Form

_____

**2)** 83,741

_____

Expanded Form

_____

**3)** 69,004

_____

Expanded Form

_____

**4)** 4,270,270

_____

Expanded Form

_____

**5)** 500,036

_____

Expanded Form

_____

**Exercises 2**   Write each number in its numerical form.

**1)**  Three thousand, two hundred eighty-six   _____

**2)**  Six hundred eighty-five thousand, thirty-one   _____

**3)**  Two hundred ninety-five thousand   _____

**4)**  Five hundred thirty-five million, two hundred sixteen thousand, fifteen

_____

## 2.  Rounding Numbers
### 1–3.  Round Whole Numbers

> a) First, look at the greatest number of the place that you want to round.
> b) Now, you should look at the digit of the **next place value**, which you want to round.
> If the digit is 5 or more, then the number being rounded increases by 1.
> If it is less than 5, then the rounding digits stays the same.

**1–4.**  Round 4,683,539 to the nearest hundred thousand.

SOLUTION

4,**6**83,539

1) Round to the hundred thousands place, which is **6**.
2) Look at the next digit **8**, which is 5 or more.
   So, the digit **6** increases by 1.

→ 4,700,000

⌐The place value that is being rounded.

Exercises 3    Round each number as followed.

**1)**    Round 71,523 to the nearest ten.

_____

**2)**    Round 1,162 to the nearest hundred.

_____

**3)**    Round 295.509 to the nearest tenth.

_____

**4)**    Round 1,726 to the nearest hundred.

_____

**5)**    Round 83,515.6392 to the nearest thousand.

_____

**6)**    Round 76,082 to the nearest ten.

_____

**7)**    Round 63,627.64 to the nearest hundred.

_____

**8)**    Round 721.009 to the nearest hundredth.

_____

**9)**    Round 45,738,843 to the nearest ten million.

_____

**10)**   Round 74,736,219 to the nearest ten million.

_____

**11)**   Round 36,356,372 to the nearest hundred thousand.

_____

**12)**   Round 50,255,643 to the nearest hundred thousand.

_____

**Exercises 4**   Round each number to the **bold** place.

1)   34,**7**04

2)   384,**9**46

3)   2**4**9,593

4)   735,5**2**0

5)   **7**46,662

6)   42,**7**49,863

7)   36,372,6**7**6

8)   263,4**8**3,394

9)   535,9**3**8

10)   49,6**4**5

11)   4**6**5,463

12)   105,46**5**

**Exercises 5**   Use the number line for the following problems.

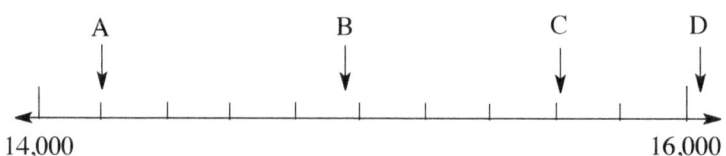

1)   What is **B** rounded to the nearest thousand?

2)   If **C** is 15,600, what other number is it closest to?

3)   Which of the letters are closest to 15,000? Then, find the place value that it is rounded to.

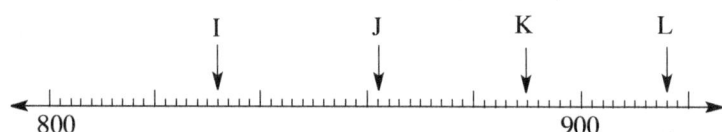

4)   If **I** is rounded to the nearest hundred, what is the number it is being rounded to?

5)   What letter is rounded to 850 if the value is rounded to the nearest ten?

6)   Which digit number is rounded if **J** is 1,000?

**SELF-TEST**

1. What is 62,466 rounded to the nearest hundred?

   **A.** 62,470                              **B.** 60,000
   **C.** 62,500                              **D.** 62,000

2. What is 90,745,735 rounded to the nearest hundred thousand?

   **A.** 90,000,000                          **B.** 91,000,000
   **C.** 90,700,000                          **D.** 90750,000

3. Which of the following is 57,000,000 that is rounded to the nearest million?

   **A.** 57,473,749                          **B.** 57,503,362
   **C.** 56,473,837                          **D.** 56,099,736

4. Which of the following could have been rounded to 800,050,000?

   **A.** 800,052,910                         **B.** 800,055,207
   **C.** 800,043,625                         **D.** 827,441,000

5. Mount Everest is 29,029 feet tall. Round Everest's height to the nearest thousand.

   **A.** 29,000 ft                           **B.** 30,000 ft
   **C.** 29,030 ft                           **D.** 20,000 ft

6. The distance between California and New York is about 2,935 miles. Round the distance to the nearest thousand.

   **A.** 2,900 miles                         **B.** 3,000 miles
   **C.** 2,940 miles                         **D.** 2,500 miles

7. What is 46,482 rounded to the nearest ten thousand?

   **A.** 46,000                              **B.** 46,500
   **C.** 50,000                              **D.** 40,000

8. What is 921,539 rounded to the nearest hundred?

   **A.** 900,000                             **B.** 920,000
   **C.** 922,000                             **D.** 921,500

* Use the number line to do Exercises **9-11**.

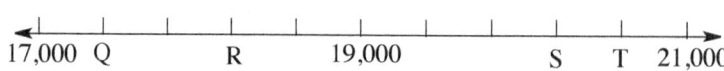

9.  Which of the following is the relative position of **R**?

     **A.**  18,600                          **B.**  17,800
     **C.**  18,200                          **D.**  17,400

10.  Which of the following represents the relative position of 19,600?

     **A. Q**                                **B. R**
     **C. S**                                **D. T**

11.  Which of the following can be rounded to 20,000 by the nearest thousand?

     **A.** Q and R                          **B.** R and S
     **C.** S and T                          **D.** T

* Use the number line for Exercises **12-14**.

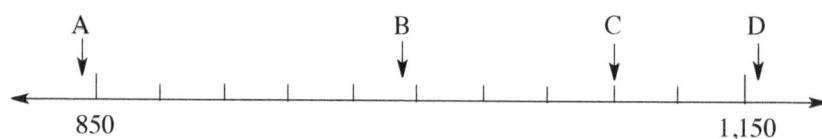

12.  What is the relative position of B?

     **A.**  990                             **B.**  1,000
     **C.**  980                             **D.**  970

13.  Which of the following represents the relative position of 1,120?

     **A.** A                                **B.** B
     **C.** C                                **D.** D

14.  Which of the following can be rounded to 800 by the nearest hundred?

     **A.** A                                **B.** A and B
     **C.** A, B, and C                      **D.** A, B, C, and D

### 3.  Estimating Sums and Differences

**1–5.**  Estimate the sum and difference.

| | | |
|---|---|---|
| **56,**537<br>**+ 74,**875 | i) The place value of ten thousand is the greatest digit.<br>ii) Round to the nearest ten thousand.<br>iii) Add. | **60,000**<br>**+ 70,000**<br>130,000 |

So, the estimated sum of 56,537 + 74,875 is about 130,000.

| | | |
|---|---|---|
| **27,**548<br>**– 14,**731 | i) The place value of ten thousand is the greatest digit.<br>ii) Round to the nearest ten thousand.<br>iii) Subtract. | **30,000**<br>**– 10,000**<br>20,000 |

So, the estimated difference of 627,548 and 84,731 is about 500,000.

**Exercises 6**   Estimate the sum and difference. Round each answer to the nearest ten thousand.

**1)**   374,763 + 736,349          **2)**   66,403 – 25,821

**3)**   49,053 – 18,437          **4)**   46,538 – 25,937

**5)**   946,538 – 275,937          **6)**   45,374 – 27,743

**7)**   100,837 + 274,439          **8)**   283,453 – 146,726

**9)**   467 – 26          **10)**   84,493 + 63,600 + 834,357

**11)**   56,494 – 5,937          **12)**   46,736 – 6,439

**Exercises 7**   Find the exact sum or difference.

1)   $203,048 + 183,209$

_____

2)   $280,063 - 195,565$

_____

3)   $48,004 - 25,416$

_____

4)   $35,355 + 68,607 + 743,043$

_____

5)   $562,839 - 97,898$

_____

6)   $210,024 - 167,465$

_____

7)
$$\begin{array}{r} 76,692 \\ - \ 32,635 \\ \hline \end{array}$$

8)
$$\begin{array}{r} 71,537 \\ 45,071 \\ + \ 82,830 \\ \hline \end{array}$$

9)
$$\begin{array}{r} 64,738 \\ 19,693 \\ + \ 68,302 \\ \hline \end{array}$$

**Exercises 8**   Find the value of $\Delta$.

1)   $8,021 + \Delta = 10,783$

_____

2)   $3,706 + \Delta = 4,245$

_____

3)   $\Delta + 6,243 = 9,304$

_____

4)   $\Delta - 3,247 = 4,200$

_____

5)   $15,055 - \Delta = 9,503$

_____

6)   $\Delta + 21,561 = 35,933$

_____

7)   $18,242 + \Delta = 35,769 - 8,055$

_____

8)   $\Delta - 4,751 = 9,163 - 2,034$

_____

9)   $8,457 + \Delta = 10,606 - 1,350$

_____

10)   $\Delta + 2,363 = 9,051 - 2,345$

_____

### 4.  Estimating Products and Quotients

**1–6.**  Estimate the product.

SOLUTION

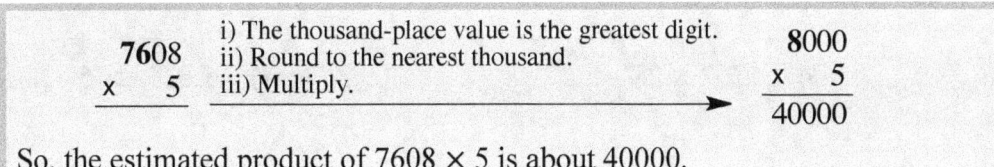

7608
x    5

i) The thousand-place value is the greatest digit.
ii) Round to the nearest thousand.
iii) Multiply.

8000
x    5
40000

So, the estimated product of 7608 × 5 is about 40000.

**1–7.**  Estimate the product.

SOLUTION

448
x 47

i) The hundred-place value is the greatest digit.
ii) Round to the nearest hundred.
iii) Round to the nearest ten.

400
x 50
20000

So, the estimated product of 448 × 47 is about 20000.

**1–8.**  Estimate the quotient.

$$2{,}947 \div 6$$

SOLUTION

i) Dividing by the thousands place is impossible.
   So skip it and divide the hundreds place value. The dividend of 29 can be
   divided.
ii) Put in zero for the rest of the digits.

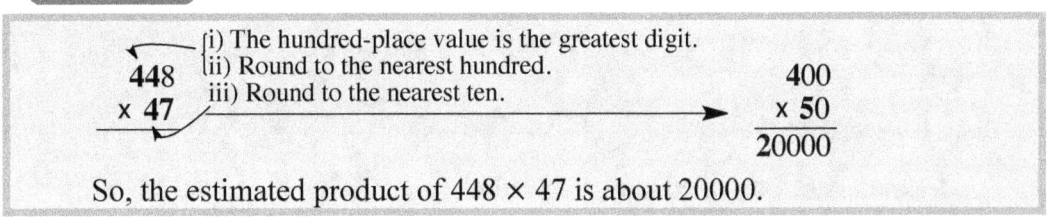

6)2947

i) The thousand-place value is the greatest digit,
ii) Dividing by the thousands place is impossible
(2/6). So divide the hundreds place(29/6). The
dividend of 29 can be divided by numbers
between 24 and 30.

iii) Replace the other digits with 0.

400         500
6)2947  or  6)2947

6 x 4 = 24
6 x 5 = 30

So, the estimated quotient of 2947 ÷ 6 is about 400 and 500.

**Exercises 9**   Estimate the product or quotient.

**1)**   $549 \times 2$

**2)**   $283 \div 7$

_____

**3)**   $267 \times 28$

**4)**   $6269 \times 5$

_____

**5)**   $254 \div 6$

**6)**   $925 \div 12$

_____

**7)**   $26,507 \div 6$

**8)**   $552 \div 8$

_____


**Exercises 10**   Estimate the product or quotient.

**1)**   $32,628 \times 27$

**2)**   $552 \div 16$

_____

**3)**   $75,072 \times 44$

**4)**   $2,492 \div 54$

_____

**5)**   $857 \div 28$

**6)**   $6,253 \times 13$

_____

**7)**
$$\begin{array}{r} 748 \\ \times\ \ 3 \\ \hline \end{array}$$

**8)**
$$\begin{array}{r} 329 \\ \times\ 31 \\ \hline \end{array}$$

**9)**
$$2\overline{)360}$$

**10)**
$$52\overline{)3729}$$

**11)**
$$\begin{array}{r} 273 \\ \times\ 59 \\ \hline \end{array}$$

**12)**
$$\begin{array}{r} 6509 \\ \times\ \ 82 \\ \hline \end{array}$$

# * Solving Problems

**Exercises 11**    Solve each problem using the given information.

1) On Sunday, 1,043 people went to the mall. The next day, 516 more people came than before. Estimate how many people total came to the mall on Monday.

2) Mount Everest is 29,029 feet tall while Mount McKinley is $\frac{2}{3}$ times taller than Mount Everest. Estimate how much taller Mount Everest is than Mount McKinley.

3) The Addams family recycled 873 plastic bottles last month. This month, they recycled half as many bottles than they did last month. Estimate how many more bottles they recycled this month.

SELF-TEST

1. Which of the following is the word form for 400,007,030?

    **A.** Four million, seven thousand, thirty
    **B.** Four million, seven thousand, and three
    **C.** Four hundred million, seven thousand, and thirty
    **D.** Four hundred million, seven hundred, and thirty

2. Which of the following is the standard form of ten million, five hundred, five thousand, and six?

    **A.** 10,505,006           **B.** 10,550,006
    **C.** 10,055,600           **D.** 10,055,006

3. Which of the following is the expanded form of 600,030,010?

    **A.** 600,000,000 + 30,000 + 100      **B.** 600,000,000 + 300,000 + 10
    **C.** 600,000,000 +300,000 + 100      **D.** 600,000,000 ⊦ 30,000 + 10

4. Which of the following is the standard form of 300,000,000 + 1,000,000 + 100,000 + 4,000 + 1?

    **A.** 310,104,001           **B.** 301,104,001
    **C.** 301,014,001           **D.** 301,140,000

5. Which of the following is the best estimated form of "73,647 – 4,450"?

    **A.** 73,000 – 4,000           **B.** 70,000 – 4,000
    **C.** 70,000 – 4,500           **D.** 74,000 – 4,000

6. What is the best estimated sum of "192,059 + 749,026"?

    **A.** 1,000,000           **B.** 900,000
    **C.** 1,200,000           **D.** 1,100,000

7. What is the best estimated product of "2,398 × 25"?

    **A.** 90,000           **B.** 40,000
    **C.** 50,000           **D.** 60,000

8. What is the best estimated quotient of "18,609 ÷ 84"?

    **A.** 250           **B.** 200
    **C.** 225           **D.** 222

9. What is the best estimated quotient of "63,046 ÷ 46"?

    **A.** 2500           **B.** 1000 or 2000
    **C.** 1370           **D.** 1340

10. The bakery is baking loaves of banana bread. They use 18,725 bunches of banana per month. If there are 8 bananas to a bunch, how many bananas do they have? Estimate to the nearest ten thousand.

    **A.** 1,870,000           **B.** 1,000,000
    **C.** 200,000           **D.** 160,000

## 5.  Reviewing Addition, Subtraction, Multiplication, and Division with Whole Numbers.

**Exercises 12**    Add each expression.

1)    $361,218 + 26,540 + 237,333$          2)    $651,730 + 74,361$

3)    $472,054 + 15,763 + 187,091$          4)    $246,284 + 172,684 + 739,631$

5)    $(354,751 + 2,401) + 376,432$          6)    $253,752 + 752,835 + 26,246$

7)
$$\begin{array}{r} 52,367 \\ 68,256 \\ +\ 47,873 \\ \hline \end{array}$$

8)
$$\begin{array}{r} 64,398 \\ 35,948 \\ +\ 245,977 \\ \hline \end{array}$$

9)
$$\begin{array}{r} 847,493 \\ 46,033 \\ +\ 473,029 \\ \hline \end{array}$$

**Exercises 13**    Find the value of Δ.

1)    $\Delta + 24,035 = 40,638$          2)    $\Delta + 11,247 = 18,208$

3)    $52,421 + \Delta = 68,073$          4)    $\Delta + 31,301 = 55,933$

5)    $25,770 + \Delta = 51,322 - 14,055$          6)    $\Delta + 19,751 = 46,163 - 12,065$

7)    $3,056 + \Delta = 21,355 - 11,350$          8)    $\Delta + 4,006 = 15,000 - 6,522$

**Exercises 14**    Subtract each expression.

1)    543,052 – 276,163

2)    631,284 – 242,928

3)    365,751 – 97,482

4)    648,272 – 459,325

5)    203,380 – 136,687

6)    211,132 – 75,489

7)
$$\begin{array}{r} 95,036 \\ -\ 27,457 \\ \hline \end{array}$$

8)
$$\begin{array}{r} 231,031 \\ -\ 57,307 \\ \hline \end{array}$$

9)
$$\begin{array}{r} 533,548 \\ -\ 55,759 \\ \hline \end{array}$$

10)
$$\begin{array}{r} 725,211 \\ -\ 47,313 \\ \hline \end{array}$$

11)
$$\begin{array}{r} 415,002 \\ -\ 64,942 \\ \hline \end{array}$$

12)
$$\begin{array}{r} 626,442 \\ -\ 356,479 \\ \hline \end{array}$$

**Exercises 15**    Find the value of $\Delta$.

1)    $53,841 - \Delta = 34,008$

2)    $69,463 - \Delta = 48,214$

3)    $\Delta - 32,917 = 56,109$

4)    $\Delta - 15,258 = 2,279$

5)    $43,453 - \Delta = 24,337$

6)    $\Delta - 31,013 = 69,255$

7)    $67,033 - \Delta = 12,447 + 9,038$

8)    $\Delta - 22,643 = 54,236 - 12,668$

**Exercises 16**   Multiply.

1)   $76,905 \times 7$                    2)   $45,044 \times 8$

_____                    _____

3)   $65,683 \times 9$                    4)   $84,007 \times 5$

_____                    _____

5)   $30,785 \times 6$                    6)   $46,932 \times 9$

_____                    _____

7)   $36,679 \times 8$                    8)   $90,400 \times 8$

_____                    _____

9)   $70,008 \times 8$                    10)  $86,926 \times 3$

_____                    _____

**Exercises 17**   Find the value of Δ.

1)   $78 \times \Delta = 702$                    2)   $21 \times \Delta = 168$

_____                    _____

3)   $\Delta \times 7 = 371$                    4)   $\Delta \times 9 = 612$

_____                    _____

5)   $47 \times \Delta = 376$                    6)   $\Delta \times 3 = 2,160$

_____                    _____

7)   $55 \times \Delta = 330$                    8)   $\Delta \times 24 = 216$

_____                    _____

**Exercises 18**    Multiply.

1)    $23 \times 247$                              2)    $45 \times 283$

_____          _____

3)    $63 \times 803$                              4)    $847 \times 25$

_____          _____

5)    $305 \times 53$                              6)    $462 \times 82$

_____          _____

7)    $36 \times 182$                              8)    $904 \times 20$

_____          _____

9)                          10)                          11)
$$\begin{array}{r} 328 \\ \times\ 21 \\ \hline \end{array} \qquad \begin{array}{r} 36 \\ \times\ 978 \\ \hline \end{array} \qquad \begin{array}{r} 675 \\ \times\ 74 \\ \hline \end{array}$$

**Exercises 19**    Find the value of Δ.

1)    $41 \times \Delta = 1{,}312$                  2)    $18 \times \Delta = 810$

_____          _____

3)    $\Delta \times 16 = 1{,}376$                  4)    $\Delta \times 21 = 1{,}008$

_____          _____

5)    $74 \times \Delta = 888$                      6)    $\Delta \times 49 = 1{,}470$

_____          _____

7)    $15 \times \Delta = 420$                      8)    $\Delta \times 24 = 816$

_____          _____

**Exercises 20**   Divide and round to the nearest whole number.

1)   $1,909 \div 23$                          2)   $900 \div 20$

_____                    _____

3)   $1,768 \div 52$                          4)   $176 \div 22$

_____                    _____

5)   $533 \div 41$                            6)   $744 \div 24$

_____                    _____

7)   $2,937 \div 43$                          8)   $8,115 \div 78$

_____                    _____

9)                          10)                          11)

$18\overline{)1363}$              $27\overline{)1735}$              $84\overline{)4957}$

**Exercises 21**   Find the value of Δ.

1)   $96 \div \Delta = 12$                       2)   $378 \div \Delta = 63$

_____                    _____

3)   $\Delta \div 25 = 32$                       4)   $\Delta \div 13 = 43$

_____                    _____

5)   $216 \div \Delta = 24$                      6)   $\Delta \div 18 = 21$

_____                    _____

7)   $624 \div \Delta = 39$                      8)   $\Delta \div 54 = 62$

_____                    _____

**Exercises 22**   Divide and find the remainder.

1)   $81,109 \div 7$                          2)   $38,584 \div 6$

_____                    _____

3)   $71,708 \div 49$                         4)   $17,558 \div 23$

_____                    _____

5)   $53,073 \div 28$                         6)   $74,834 \div 88$

_____                    _____

7)   $29,937 \div 45$                         8)   $98,005 \div 39$

_____                    _____

9)                           10)                          11)
   $36\overline{)3472}$           $42\overline{)2402}$          $63\overline{)3095}$

**Exercises 23**   Find the value of Δ.

1)   $5,568 \div \Delta = 87$                    2)   $1,504 \div \Delta = 16$

_____                    _____

3)   $\Delta \div 76 = 52$                        4)   $\Delta \div 19 = 95$

_____                    _____

5)   $2,052 \div \Delta = 54$                    6)   $\Delta \div 82 = 48$

_____                    _____

7)   $4,161 \div \Delta = 73$                    8)   $\Delta \div 47 = 76$

_____                    _____

**Exercises 24**   Find the value of each box.

**1)**

$$24\overline{)\phantom{00000}}\;403r20$$

**2)**

$$43\overline{)\phantom{00000}}\;69r42$$

**3)**

$$18\overline{)\phantom{00000}}\;133r17$$

**4)**

$$82\overline{)\phantom{00000}}\;78r69$$

**5)**

$$29\overline{)\phantom{00000}}\;371r25$$

**6)**

$$12\overline{)\phantom{00000}}\;986r9$$

## * Solving Problems

**Exercises 25**   Solve each problem using the given information.

**1)** The ice cream store sells vanilla, strawberry and chocolate ice cream. At the end of the day, the owner found that 7 times as many chocolate ice creams were sold than strawberry ice cream and 3 times as many vanilla ice creams were sold than chocolate. If the owner sold 355 vanilla ice creams, how many strawberry ice creams did the owner sell? Show your work.

**2)** A bakery store used eggs to bake chocolate chip cookies and cakes. The chocolate chip cookies need 8 times as many eggs than the cakes. If the baker used 648 eggs to make chocolate chip cookies, how many eggs did the baker use to make the cakes? Show your work.

**3)** A plane is depositing fire retardant powder on a forest fire. On the first run, the plane dumps powder over 42 acres. On the third run, the plane dumps powder half as much than the first run, but 3 times more than the second run. How many acres did the plane dump in the second run? How many acres did the plane dump on the third run?

4)  Farmer Carter grows cabbages every year. In 2012, he grew 4 times as many cabbages in 2010. In 2013, he grew half of the amount of cabbages he had grown in 2010. If he grew 26,524 cabbages in 2010, how many cabbages did he grow in 2012? How many cabbages did he grow in 2013? Show your work.

5)  Three tables are set out for a buffet. On each table are two plates of shrimp. Both the plates on the first table have exactly 248 pieces of shrimp. The second table has twice as much shrimp on each plate than the first table and the third table has 3 times as less shrimp than the sum of the first table and the second table. How many pieces of shrimp are there on the second table? How many pieces of shrimp are there on the third table? Show your work.

6)  Lucy has $1,023 in her bank account. She deposited money in her account for two weeks. On the first week, she deposited twice as much than she had the previous week, and on the second week she deposited 3 times less than what she had the previous week. If she deposited $12 in the second week, how much money does Lucy have in her bank account in the first week? How much money does Lucy have in her bank account in the first and second weeks? Show your work.

7)  Daniel is buying shirts. He spends $23 for a shirt and then buys two more shirts that both cost $2 less than the first shirt. If Daniel initially has $84 in total, how much money does he have left? Show your work.

**SELF-TEST**

1. What is the sum of the following expression?
   48,938 + 15,363

   **A.** 70,000                    **B.** 60,000
   **C.** 54,301                    **D.** 64,301

2. What is the sum of the following expression?
   2,403,239 + 86,809

   **A.** 11,084,139                **B.** 2,490,048
   **C.** 2,500,000                 **D.** 2,490,000

3. What is the difference of the following expression?
   74,025 − 15,026

   **A.** 58,999                    **B.** 68,999
   **C.** 58,765                    **D.** 59,000

4. At Halloween, Mr. Myers gives away 7 pieces of candy to each trick-or-treater. If 57 trick-or-treaters came to his door, how many pieces of candy did he give away?

   **A.** 420                       **B.** 50
   **C.** 64                        **D.** 399

5. A bakery has 43 bags of bagels. Each bag contains 12 bagels. How many bagels does the bakery have?

   **A.** 55                        **B.** 31
   **C.** 516                       **D.** 3.6

6. A tray can hold 65 eggs. If there are 1,495 eggs, how many trays will be needed?

   **A.** 1,560                     **B.** 1,430
   **C.** 97,175                    **D.** 23

7. At an art exhibition, nails are needed in order to hang up the paintings. If a painting needs 14 nails to stay on the wall and there are 168 paintings, how many nails are needed in total? Round to the nearest whole number.

A. 12                                        B. 182
C. 154                                       D. 2,352

8.  What is the product of the following expression?
    23,950 × 7

    A. 3,421                                 B. 167,650
    C. 23,957                                D. 257

9.  What is the difference of the following expression?
    10,370 − 728

    A. 9,642                                 B. 642
    C. 9,800                                 D. 9,000

10.  What is the quotient of the following expression?
     635,401 ÷ 37

     A. 15,000                               B. 14,000
     C. 17,173                               D. 15,885

11.  What is the value that fits in the box?
     $537,733 + \boxed{\phantom{xxx}} = 903,304$

     A. 365,571                              B. 1,441,037
     C. −365,571                             D. 1.68

12.  What is the value of Δ for the expression below?
     $4,617 ÷ Δ = 243$

     A. 19                                   B. 1,121,931
     C. 4,374                                D. 4,860

13.  What is the value of Δ for the expression below?
     $10,321 − Δ = 3,726$

     A. 6,595                                B. 14,047
     C. -6,596                               D. 2.77

14. Which of the following expressions exemplifies the relationship of the equation with the remainder below?
$$35,281 \div 59 = 597r58$$

  **A.** $(597 \times 59) \times 58$       **B.** $(597 \times 59) + 58$
  **C.** $(597 + 58) \times 59$       **D.** $(59 + 597) \times 58$

15. Which of the following equations exemplifies the relationship of the equation with the remainder below?
$$3,325 \div 52 = 63r49$$

  **A.** $3,325 \div 52 = 63 + 49$     **B.** $63 + \dfrac{49}{52} = 3,325 \div 52$
  **C.** $3,325 = (63 + 52) \times 49$     **D.** $(63 + 49) \times 52 = 3,325$

16. Which of the following does not match the equation below?
$$2,695 \div 31 = 86r29$$

  **A.** $(2,695 - 29) \div 31 = 86$     **B.** $86 + \dfrac{29}{31} = 2,695 \div 31$
  **C.** $2,695 = (86 \times 31) + 29$     **D.** $(2,695 + 29) \div 31 = 86$

17. Find the missing operation sign represented by $\Delta$.
$$288 \, \Delta \, 24 + 7,018 = 7,030$$

  **A.** $\div$          **B.** $\times$
  **C.** $+$          **D.** $-$

18. Find the missing operation sign represented by $\Delta$.
$$79,637 \, \Delta \, 10,873 = 90,510$$

  **A.** $\div$          **B.** $\times$
  **C.** $+$          **D.** $-$

## 6.  Order of Operations

**1-9.** Order of operations

---

a) Order of operations is always done in order: 1) First solve within the parentheses, 2) Multiply and divide, 3) Then add and subtract. Solve from left to right to when adding and subtracting or multiplying and dividing in an expression.

b) If there are no parentheses or grouping symbols, multiplying and dividing is always done before adding and subtracting.

---

**Exercises 26**   Write each expression in its word form. Do not solve the expression.

**1)**  $(3 - 4) \times (3 + 4)$

_____

**2)** $4 + (3 \div 2) - 1$

_____

**3)** $(8 \div 4) - 3$

_____

**4)** $(6 \times 7) + 8$

_____

**5)** $(7 - 3) \times 4 - 2$

_____


**Exercises 27**   Write each expression in its standard form. Do not solve the expression.

**1)**  Multiply the difference of five and three by six_____

**2)** Multiply the sum of nine and two by two, then add four. _____

**3)** The quotient of ten and two decreased by two.  _____

**4)** Add the sum of eight and two with the quotient of nine and three.

_____

**5)** Subtract the sum of two and eight from the product of eight and three.

_____

**Exercises 28**    Solve each order of operations.

1)    $(25 - 6) + (820 - 734)$          2)    $422 - (301 - 19) + 630$

_____            _____

3)    $20 + 5 \times 3$                        4)    $4 \times 7 - 4 \div 2$

_____            _____

5)    $3 \times (32 - 24) + 6 \div 3$            6)    $5 + 2 \times 3 - 4$

_____            _____

7)    $35 + 8 - 3 \times 9$                  8)    $56 - 24 \div 8$

_____            _____

9)    $18 + 54 \div 9$                  10)    $72 \div 6 - 6$

_____            _____


**Exercises 29**    Solve each order of operations.

1)    $7 \times 9 - 2 \times 0.5$                2)    $5 - 4 + 0.75 \times 12$

_____            _____

3)    $0.57 \times 7 + 9 \div 3$              4)    $95 \div 5 - \frac{1}{4} \times (7 - 3)$

_____            _____

5)    $(22 + 4) \div 2$                    6)    $18 - (9 \div 3) \times \frac{2}{3}$

_____            _____

7)    $23 - (3 \times 2\frac{1}{3}) + 2$            8)    $(19 - 15) \times [\frac{3}{5}(3 + 2)]$

_____            _____

## *Solving Problems

Exercises 30    Solve each problem using the given information.

1) Lucy needs to sell 120 cups of lemonade for the school fundraiser. There are equal numbers of yellow, blue and red cups. She sold 5 more yellow cups than blue cups, and 2 more red cups than yellow cups. If she sold 30 blue cups, how many cups does she have left over in total? How many blue cups does she have left?

2) At the school festival, the PTA is preparing hot dogs and hamburgers. The number of hamburgers eaten is half the amount of hot dogs eaten. If students eat 1,732 hot dogs, how many hamburgers did they eat?

**SELF-TEST**

1.  Which of the following is the value of the following expression?
    Eleven times the sum of one and six

    **A.** 88                                    **B.** 77
    **C.** 66                                    **D.** 55

2.  Which of the following statement best describes the expression below in standard form?
    Add the quotient of ten and five with the product of three and two

    **A.** $(10 + 5) \div (3 \times 2)$          **B.** $(10 \times 5) \div (3 + 2)$
    **C.** $(10 \div 5) + (3 \times 2)$          **D.** $(10 \div 5) \times (3 + 2)$

3.  Which of the following is not correct?
    $(9 + 6) \div (3 - 2)$

    **A.** The quotient between the sum of nine and six and the difference of three and two
    **B.** The sum of nine and six divided by the difference of three and two
    **C.** The sum of nine and six divided by three decreased by two
    **D.** Add nine and six divided by three increased by two

4. Which of the following is the value of the expression below?
   Subtract the quotient of thirty-eight and two from the product of ten and one.

   A. 8                                   B. 9
   C. 10                                  D. 29

6. Find the value of the expression below.
$$3 + 8 \times \frac{1}{4} - 6 \div 2$$

   A. 1                                   B. 2
   C. 3                                   D. 4

7. Find the missing operation sign represented by Δ.
$$20 \, \Delta \, \frac{3}{5} + (7 - 3) \div 2 = 14$$

   A. ÷                                   B. ×
   C. +                                   D. −

9. James is driving his car at 45 miles/hour while Kim is driving $1\frac{2}{5}$ times faster than James. How much faster is Kim going than James?

   A. 112 miles/hour                      B. 22 miles/hour
   C. 1.5 miles/hour                      D. 18 miles/hour

10. The population of Elizabeth, New York is 125,660 people while the population of Anaheim, California is $2\frac{7}{10}$ times as many as Elizabeth, New York. How many more people does Anaheim have than Elizabeth? Round to the nearest whole number.

   A. 467,021                             B. 215,701
   C. 213,622                             D. 216,701

11. Andrew is counting his coins. He has organized his coins so that there are 32 coins in each bag. If Andrew has $7\frac{1}{5}$ times as many coins than bags, how many coins does he have? Round to the nearest whole number.

   A. 7 coins                             B. 192 coins
   C. 252 coins                           D. 7,168 coins

**12.** Reorder the following expression so that it follows the order of operations.

$$\frac{3}{5} \times 25 - 25 \div 5 + 74$$

**A.** $\frac{3}{5} \times (25 - 25) \div 5 + 74$       **B.** $\frac{3}{5} \times (25 - 25) \div (5 + 74)$

**C.** $(\frac{3}{5} \times 25) - (25 \div 5) + 74$       **D.** $(\frac{3}{5} \times 25) - 25 \div (5 + 74)$

**13.** Find the value of the expression below.

$$4 + 10 \times \frac{3}{5} - 4 \div 2$$

**A.** 8                 **B.** 40

**C.** 20                **D.** 54

**14.** Use the expression below.

$$12 \times \frac{3}{5} \div 6 - 5$$

Which of the following correctly follows the order of operations?

**A.** $12 \times [(\frac{3}{5} \div 6) - 5]$       **B.** $[12 \times (\frac{3}{5} \div 6)] - 5$

**C.** $[(12 \times \frac{3}{5}) \div 6] - 5$       **D.** $(12 \times \frac{3}{5}) \div (6 - 5)$

**15.** Which of the following is correct?

    **A.** $625,107 < 625,097$
    **B.** $837,000 > 840,000$
    **C.** 100,101 is less than 100,110.
    **D.** 1,000,000 is greater than one million.

**16.** Which of the following is incorrect?

    **A.** $372,374 \neq$ three hundred seventy-two thousand and three hundred seventy-four
    **B.** 2,603,002 is greater than two millions.
    **C.** $109,820 = 100,000 + 9,000 + 800 + 20$
    **D.** ten thousand, twenty-three $= 10,023$

# CHAPTER 2
# Patterns, Function, and Algebra

In this chapter, you will identify number sense based on the number system, algebraic expression, and order of operations. Also you will learn about graphing functions, plotting the coordinate, relating function tables, and finding the equations of function tables.

## 1. Solving Equations with One Variable: Addition and Subtraction

**2–1.** Solve the equation.

$$9 \text{ kg} + x \text{ kg} = 23 \text{ kg}$$

**SOLUTION**

a) First, look at the equation and the figure. The $x$ represents the unknown value, which is called the variable.

$$9 + x = 23$$

| 1 kg | 2 kg | 9 kg | 12 kg | 14 kg | 15 kg | 18 kg | 20 kg | 23 kg |

$x$ kg
9 kg ———————————— 23 kg

The equation can be compared to a balance, which has exactly the same weight on both sides.

b) Take 9 kg from both sides and they are still balanced on both sides.

| $9 + x = 23$ | Original equation |
| $(-9) + 9 + x = 23 + (-9)$ | Subtract (**9**) from both sides. |
| $x = 14$ | Simplify. $(-9) + 9 = 0$, $23 + (-9) = 14$ |

$x$ kg ———————————— 14 kg

9 kg                                    9 kg

9 kg were removed from each side. The left side of the balance remain only $x$ kg and then right side of the balance is 14 kg. Therefore, the left side of $x$ kg on the

balance should be 14 kg.

Check the solution.

| 9 kg | + | x | = | 23 kg |

| 9 kg | + | 14 kg | = | 23 kg |

| 23 kg | = | 23 kg |

**2–2.** Solve the equation.

$$5(4 + x) = 40$$

**SOLUTION**

For solving for $x$, you should divide 5 from each side, which cancels out 5 on the left side. 8 will be on the right side. Next, subtract 4 from each side. Now $x$ can be solved for.

$5(4 + x) = 40$      Original equation

$\frac{1}{5} \times 5(4 + x) = 40 \times \frac{1}{5}$      Divide each side by 5.

$4 + x = 8$      Simplify. $\frac{1}{5} \times 5 = 1, 40 \times \frac{1}{5} = 8$

$4 - 4 + x = 8 - 4$      Subtract 4 from both sides.

$x = 4$

So, the value of $x$ is 4.

Check the solution:

$5(4 + x) = 40$      Original equation

$5(4 + 4) = 40$      Substitute $x$ with 4.

$40 = 40$      Simplify. It means $x = 4$ is true.

**2–3.** Find the value.

If $x = 2$, what is the value of $2x + 2$?

**SOLUTION**

$2x + 2$      Given expression

$(2 \times 2 + 2)$      Substitute 2 for $x$.

$6$      Simplify. $2 \times 2 = 4$

So, the value of $2x + 2$ is 6.

**Exercises 1**    Solve each equation.

1)    $c + 5 = 28$                              2)    $9 + y = 37$

_____                    _____

3)    $11 + 3x = 17$                           4)    $5n + 1 = 6$

_____                    _____

5)    $2(c + 3) = 6$                           6)    $12 + 2y = 16$

_____                    _____

7)    $1 + 2y = 3$                             8)    $3(x + 2) = 12$

_____                    _____

9)    $2(x + 1) = 14$                          10)    $5 + 2y = 7$

_____                    _____


**Exercises 2**    Find the value of each variable.

1)  If $n = 4$, find the value of $19 + n$.        2)  If $x = 12$, find the value of $x + 3$.

_____                    _____

3)  If $x = 6$, find the value of $12 + 2x$.      4)  If $y = 2$, find the value of $3y + 4$.

_____                    _____

5)  If $x = 5$, find the value of $4 + 3x$.       6)  If $x = 4$, find the value of $9 - 2x + 1$.

_____                    _____

7)  If $y = 10$, find the value of $2y + (4 - 1)$.   8)  If $x = 5$, find the value of $3 + 4x$.

_____                    _____

9)  If $x + 6 = 8$, find the value of $6 + x$.    10)  If $x + 3 = 5$, find the value of $2(3 + x)$.

_____                    _____

**Exercises 3**    Solve each equation.

1)    $0.05 + x = 0.75$                              2)    $n + 4.9 = 12$

_____                          _____

3)    $c + 5.4 = 54$                                   4)    $2 + y = 7.25$

_____                          _____

5)    $9.2 + x = 18$                                    6)    $n + 7 = 16.55$

_____                          _____

7)    $c + 3.05 = 9.10$                               8)    $2.4 + 2y = 6.8$

_____                          _____

9)    $2(x + 1.8) = 14.8$                            10)    $3(x + 2.5) = 17.1$

_____                          _____


**Exercises 4**    Find the value of each expression.

1)  If $1 + x = 10$, find the value of $x(1 + x)$.        2)  If $2 + x = 5$, find the value of $2(x + 2)$.

_____                          _____

3)  If $n + 1 = 3$, find the value of $1 + 2n$.        4)  If $y = 2$, find the value of $3y + 4$.

_____                          _____

5)  If $x = 0.5$, find the value of $2(1 + x)$.        6)  If $c = 2.3$, find the value of $0.3 - c + 4.6$.

_____                          _____

7)  If $y + 5 = 7$, find the value of $2(y - 7)$.        8)  If $3 + x = 9$, find the value of $4 + x - 1$.

_____                          _____

9)  If $1 + \frac{1}{2}x = 2$, find the value of $1 + \frac{1}{2}x$.        10)  If $x = \frac{1}{2}$, find the value of $2(1 + 2x)$.

_____                          _____

**2–4.** Solve the equation.

$$k - 13 = 24$$

SOLUTION

For solving $k$, you should add 13 to each side and then cancel out 13 on the left side. So only $k$ will remain and then 37 will be left on the other side of the equation.

You look at the equation of $k - 13 = 24$, where $k$ is the variable.

| | |
|---|---|
| $k - 13 + \mathbf{13} = 24 + \mathbf{13}$ | Add 13 to both sides. |
| $k = 37$ | Simplify. $13 - 13 = 0$, $24 + 13 = 37$ |

So, the value of $k$ is 37.

Check the solution:

| | |
|---|---|
| $k - 13 = 24$ | Original equation |
| $37 - 13 = 24$ | Substitute $k$ with 37. |
| $24 = 24$ | Simplify. $37 - 13 = 24$ |

**2–5.** Find the value.

If $x - 2 = 4$, what is the value of $2(x - 2)$?

SOLUTION

You can solve the problem using two methods.
i) Find the value of $x$ and then find the value of the expression.

| | |
|---|---|
| $x - 2 = 4$ | Given |
| $x - 2 + 2 = 4 + 2$ | Add 2 to both sides. |
| $x = 6$ | Simplify. $2 - 2 = 0$, $2 + 2 = 4$ |

Now, find the value of $2(x - 2)$.

| | |
|---|---|
| $2(6 - 2)$ | Substitute $x$ with 6. |
| 4 | Simplify. $(6 - 2) = 4$, $4 \times 2 = 8$ |

So, the value of $2(x - 2)$ is 8.

ii) Substitute 4 for $x - 2$.

| | |
|---|---|
| $2(x - 2)$ | Given |
| $2(4)$ | Substitute 4 for $x - 2$. |
| 8 | Simplify |

So, the value of $2(x - 2)$ is 8.

You can solve the problem both ways and the answer will be identical.

**Exercises 5**    Find the value of each variable.

1)    $2(2-9)+23=14+k-15$          2)    $13-a=19+3(2-11)$

_____                              _____

3)    $20+2(7+c)=92-11+9$          4)    $3(n+6)-14=132-26$

_____                              _____

5)    $2(x-5)+13=-10-22+89$        6)    $8+9-1=6+5(n-2)$

_____                              _____

**Exercises 6**    Solve each equation.

1)    $c-7=21$                          2)    $19-y=15$

_____                              _____

3)    $10-3x=7$                         4)    $2n-1=9$

_____                              _____

5)    $2(c-9)=24$                       6)    $2(y-3)=18$

_____                              _____

7)    $8-3x=-25$                        8)    $3(c-5)=9$

_____                              _____

9)    $3(x-10)=3$                       10)    $2n-1=53$

_____                              _____

11)                          12)                          13)

$$\begin{array}{r} 26 \\ -\ y \\ \hline 17 \end{array}$$          $$\begin{array}{r} 61 \\ -\ n \\ \hline 22 \end{array}$$          $$\begin{array}{r} x \\ -\ 8 \\ \hline 29 \end{array}$$

**Exercises 7**    Solve each equation.

1)    $4.5 - x = 8$                              2)    $n - 2.05 = 2$

_____                          _____

3)    $c - 9.1 = 3$                             4)    $1.4 - y = 4$

_____                          _____

5)    $c - 6 = 17.5$                            6)    $3 - y = 6.9$

_____                          _____

7)    $2(2 - x) = 4.4$                          8)    $2(c - 3.6) = 8$

_____                          _____

**Exercises 8**    Find the value of each expression.

1)  If $n = 1$, find the value of $21 - 2n$.          2) If $x = 2$, find the value of $2(x - 1)$.

_____                          _____

3)  If $x = 6$, find the value of $2x - 5$.          4)  If $y = 3$, find the value of $3(2y - 4)$.

_____                          _____

5)  If $x - 1 = 2$, find the value of $2(x - 1)$.          6)  If $c - 6 = 18$, find the value of $6 + c - 6$.

_____                          _____

7)  If $1 - y = 4$, find the value of $4(2 - y)$.          8)  If $10 - x = 5$, find the value of $2(9 - x)$.

_____                          _____

9) If $12 - \frac{1}{2}x = 8$, find the value of $(12 - \frac{1}{2}x) +$          10)  If $x = 2$, find the value of $2(6.4 - x)$.
    4.

_____                          _____

**Exercises 9**    Find the value of each variable.

1)    $2 + a = 15 - 3$

2)    $52 + x = 97 - 9$

3)    $14 - 12 = \frac{1}{2}(s - 6)$

4)    $22 - r = \frac{1}{3}(8 + 7)$

5)    $4 + 3a = 17 - 1$

6)    $28 - 2 \times 7 = 7(s - 16)$

7)    $4 + 4(12 - k) = 27 - 11$

8)    $26 - 5 = 3k + 6$

9)    $1 - a = 7 - 15$

10)    $5.2 - x = 13.1 - 19.4$

11)    $9 + 2a = 14 - 3$

12)    $3.7 + 2a = 19.9 - 3$

**SELF-TEST**

1.  Which of the following expressions represents the difference of 3 times $x$ and 7?

    **A.** $3 \div 7x$            **B.** $3(x - 7)$
    **C.** $3(x + 7)$            **D.** $3x - 7$

2.  What is the value of $x$ for the equation below?
    $$2x - 4 = 18$$

    **A.** 7            **B.** 11
    **C.** 14            **D.** 22

3. What is the value of $x$ for the equation below?
$$2(x-4) = 18$$

   **A.** 13                    **B.** 5
   **C.** 26                    **D.** 32

4. What is the value of $x$ for the equation below?
$$\frac{1}{2}(x-4) = 18$$

   **A.** 13                    **B.** 40
   **C.** 22                    **D.** 36

\* For Exercises **5-6**. Will has 28 bottles of water in his refrigerator and drinks 2 bottles. He then drinks 3 bottles every day for the next 5 days.

5. Write an equation to represent how many bottles of water are left.

   **A.** $(28 - 2) + (3 \times 5) = x$          **B.** $(28 + 2) - (3 \times 5) = x$
   **C.** $(28 + 2) - (3 \times 5) = x$          **D.** $(28 - 2) - (3 \times 5) = x$

6. How many bottles of water are left in his refrigerator?

   **A.** 11                    **B.** 8
   **C.** 22                    **D.** 19

7. Linda bought 30 hairpins at the beauty shop and gave them to each of her 5 friends. If she gave each friend an equal number of hairpins and has 5 remaining, what equation could be used to show how many hairpins she gave to her friends?

   **A.** $30 + 5x = 5$          **B.** $30 \div 5x = 5$
   **C.** $30 \times 5x = 5$          **D.** $30 - 5x = 5$

8. How many hairpins did she give to her friends?

   **A.** 10                    **B.** 15
   **C.** 25                    **D.** 5

\* For Exercises **9-10**. Cark has 15 chocolate bars. He eats 4 bars and gives to his friend 2 more bars than what he just ate.

9. Write an equation to represent how many bars are left.

    **A.** $15 = x + 4 - (4 + 2)$            **B.** $15 = x + 4 \div (4 + 2)$
    **C.** $15 = x + 4 \times (4 + 2)$          **D.** $15 = x + 4 + (4 + 2)$

10. How many bars does he have left?

    **A.** 5                  **B.** 8
    **C.** 9                  **D.** 10

* Nick borrowed 6 nonfiction books from the library. He notices that he borrowed $\frac{1}{3}$ as many nonfiction books than fiction books.

11. If he returned 2 nonfiction books to the library, how many nonfiction books does he have?

    **A.** 0                  **B.** 1
    **C.** 2                  **D.** 3

12. Which of the following expressions represents "the sum of 2 times $x$ and 2"?

    **A.** $2 \div x$              **B.** $2 \times x + 2$
    **C.** $2(2 + x)$           **D.** $2 = x$

13. If $x = 15$, what is the operation sign of $\Delta$?
$$(5 \, \Delta \, 7) - (36 \div 3) = x + 8$$

    **A.** $\div$                **B.** $\times$
    **C.** $+$                 **D.** $-$

14. What is the value of $x$ for the equation below?
$$13 + 2x = 19$$

    **A.** 1                  **B.** 2
    **C.** 3                  **D.** 4

15. What is the value of $x$ for the equation below?
$$\frac{1}{3}(1 + 2x) = 9$$

    **A.** 1                  **B.** 2
    **C.** 3                  **D.** 4

**16.** What is the value of $x$ for the equation below?

$$\frac{1}{4}(3x) = 3$$

        **A.** 1                                 **B.** 3

        **C.** 2                                 **D.** 4

\* For Exercises **17-18**. Dave puts different amounts of candy in 3 paper bags. The first bag contains 8 pieces of candy. The second box has half the amount of candy than the first bag. He does not remember how many pieces he put into the third bag.

**17.** What is the equation that represents the problem given that he has 30 total pieces of candy?

    **A.** $8 + \frac{1}{2} \times 8 + 30 = k$               **B.** $30 + k = 8 + 8 \times \frac{1}{2}$

    **C.** $8 + 8 \times \frac{1}{2} + k = 30$              **D.** $8 + 8 \times 2 - 30 = k$

**18.** How many pieces of candy does Dave put in the third bag?

        **A.** 12                               **B.** 16

        **C.** 18                               **D.** 20

**19.** Julia bought 3 different kinds of fruits in a grocery market, which are apples, pears, and oranges. She bought that 4 times as many apples than pears and $\frac{1}{2}$ as many oranges than pears. If she bought 16 apples, how many oranges did she buy?

        **A.** 1                                 **B.** 2

        **C.** 3                                 **D.** 4

**20.** What is the value of $x$ for the equation below?

$$2.47 + 2x = 8.63$$

        **A.** 3.08                              **B.** 6.16

        **C.** 11.1                              **D.** 5.55

## 2.   Solving Equations with One Variable: Multiplication and Division

**2–6.**  Solve the equation.

$$5 \times n = 55$$

**SOLUTION**

To solve for $n$, you should divide each side by 5 and then cancel out 5 on the left side. So only $n$ will remain on the left side and 11 will be on the right side.

The operation sign ($\times$) can be omitted between a number and a variable without changing the meaning.

$5 \times n = 55$  or  $5n = 55$

$\frac{1}{5} \times 5 \times n = 55 \times \frac{1}{5}$          Divide each side by 5.

$\frac{5^1}{5_1} \times n = \frac{55^{11}}{5_1}$          Simplify.

$n = 11$          Simplify with the GCF.

So, the value of $n$ for $5 \times n = 55$ is 11.

Also you can check the solution by substituting the value of $n$ in the equation.

$5 \times n = 55$          Given equation

$5 \times 11 = 55$          Substitute $n$ with 11.

$55 = 55$

This means that the value of $n$ is 11.

**2–7.**  Find the value.

If $2x = 3$, what is the value of $4x + 2$?

**SOLUTION**

You can solve the problem using two ways.

i) First, solve the equation for $2x = 3$.

$\frac{1}{2} \times (2x) = 3 \times \frac{1}{2}$  Multiply.

$x = \frac{3}{2}$          Simplify.

Now, find the value of the expression.

$4x + 2$          Given expression

$(4 \times \frac{3}{2}) + 2$   Substitute $x$ with $(\frac{3}{2})$.

$8$          Simplify. $(4 \times \frac{3}{2}) = 6$

So, the value of $4x + 2$ is 8.

ii) First, look at the equation.

$2x = 3$      Given equation

$4x = 6$      Multiply 2 on both sides

Now, find the value of the expression.

$4x + 2$          Given expression

$6 + 2$          Substitute $4x$ for 6

$8$          Simplify. $6 + 2 = 8$

So, the value of $4x + 2$ is 8.

**Exercises 10**   Solve each equation.

1)   $y \times 6 = 24$

2)   $2 \times x = 110$

3)   $z \times 8 = 96$

4)   $12n = 72$

5)   $36a = 36$

6)   $5c = 155$

7)   $5y + 14 = 349$

8)   $3x - 2 = 616$

9)   $4y + 29 = 193$

10)   $7x - 42 = 91$

**Exercises 11**   Find the value of each variable.

1)   If $n = 2$, find the value of $7 - 2n$.

2)   If $x = 5$, find the value of $2(x - 3)$.

3)   If $x = 2$, find the value of $12 - 5x$.

4)   If $y = 7$, find the value of $7y + 4$.

5)   If $x = 5$, find the value of $2(9 - x)$.

6)   If $c = 8$, find the value of $17 - 2c + 3$.

7)   If $y = 1.4$, find the value of $6(y + 4.8)$.

8)   If $x = 5$, find the value of $2(3 + 2)x$.

9)   If $4x = 8$, find the value of $12 - 2x$.

10)   If $2x = 4$, find the value of $2(6 - 2x)$.

**Exercises 12**    Solve each equation.

**1)**    $7 + 2n = 31$                                 **2)**    $6n - 9 = 123$

_____                                    _____

**3)**    $8y - 13 = 202$                             **4)**    $5 + 4x = 55$

_____                                    _____

**5)**    $7(z - 4) = 21$                               **6)**    $2.5n = 32.5$

_____                                    _____

**7)**    $6(3 + a) = 42$                               **8)**    $2(c + 1) = 8$

_____                                    _____

**9)**    $5(y - 1) = 15$                               **10)**    $3(1 + x) = 12$

_____                                    _____


**Exercises 13**    Find the value of each variable.

**1)**    If $n = 4$, find the value of $23 - 2n$.      **2)**    If $x = 8$, find the value of $x - 3$.

_____                                    _____

**3)**    If $x = 2$, find the value of $12 - 5x$.      **4)**    If $y = 0.5$, find the value of $7(4y - 1)$.

_____                                    _____

**5)**    If $3x = 2$, find the value of $2(4 - 3x)$.    **6)**    If $2c = 6$, find the value of $10 - 2c + 6$.

_____                                    _____

**7)**    If $2y = 7$, find the value of $6y - 3$.       **8)**    If $2x - 3 = 5$, find the value of $(2x - 3) - 2$.

_____                                    _____

**9)**    If $4(x - 9) = 8$, find the value of $x - 9$.    **10)**    If $7(11 + x) = 63$, find the value of $2(11 + x)$.

_____                                    _____

**2–8.** Solve the equation.

$$8 \div n = 2$$

SOLUTION

If $n$ is a divisor or a denominator in the fraction, you can solve the problem in two ways.

a) $n$ should be transferred from the left side to right side in order to solve the equation. So you should multiply each side by $n$, which cancels out the $n$ on the left side. Now the variable of $n$ is on the right side of the equation. To solve for $n$, you should divide each side by 2 to cancel out the number on the right side. The $n$ will remain on the right side and 4 on the left.

A division problem can be rewritten as a fraction problem.

$8 \div n = 2$  or  $\frac{8}{n} = 2$

$\frac{8}{n} = 2$                         Rewrite as a fraction.

$n \times \frac{8}{n} = 2 \times n$              Multiply $n$ on both sides.

$8 = 2n$                         Simplify. $\overset{1}{n} \times \frac{8}{\overset{}{n_1}} = 8$

$\frac{8}{2} = \frac{2}{2}n$                        Divide 2 from both sides.

$\frac{{}^4 8}{{}_1 2} = \frac{{}^1 2}{{}_1 2}n$                       Simplify.

$4 = n$                          Simplify with the GCF.

So, the solution of $8 \div n$ is 4.

b) Given an equation like $8 \div n = 2$, you can rewrite it so that $8 = 2n$.

$8 \div n = 2$           Original equation

$8 = 2n$              Rewrite the equation as $8 \div n = 2$.

$\frac{8}{2} = \frac{2}{2}n$            Divide 2 from both sides.

$4 = n$               Simplify.

So, both answers produced by a) and b) are identical. Therefore, both ways are valid.

c) Once you solved the equation to get $n = 4$, you can check your solution by substituting the value in the original equation.

$8 \div n = 2$  or  $\frac{8}{n} = 2$  or  $8 = 2n$

When you insert $n = 4$ in all of the equations, the solution is true.

**2–9.** Solve the equation.

$$48 \div 4x = 3$$

SOLUTION

| | |
|---|---|
| $48 \div 4x = 3$ | Original equation |
| $48 = (3)(4x)$ | Rewrite the equation as $48 \div 4x = 3$. |
| $48 = 12x$ | Simplify. $(3)(4x) = 12x$ |
| $\dfrac{48}{12} = \dfrac{12}{12}x$ | Divide each side by 12. |
| $4 = x$ | Simplify with the GCF. $\dfrac{48}{12} = 4, \dfrac{12}{12} = 1$ |

So, the value of $x$ is 4.

Check to verify the solution.

| | |
|---|---|
| $48 \div 4x = 3$ | Original equation |
| $48 \div 4(4) = 3$ | Substitute $x$ with 4. |
| $48 \div 16 = 3$ | Simplify. $4 \times 4 = 16$ |
| $3 = 3$ | Simplify. $\dfrac{48}{16} = 3$ |

This means $x = 4$ is true.

**2–10.** Find the value.

If $2 \div x = 5$, what is the value of $4(2 \div x)$?

SOLUTION

First, find the value of $x$ and then find the value of the expression. However, you can replace $2 \div x$ for 5.

| | |
|---|---|
| $4(2 \div x)$ | Given expression |
| $4(5)$ | Substitute 5 for $2 \div x$. |
| $20$ | Simplify. $4 \times 5 = 20$ |

So, the value of $4(x \div 2)$ is 20.

**Exercises 14**    Solve each equation.

1)    $x \div 3 = 3$                         2)    $48 \div n = 2$

_____                        _____

3)    $2x \div 8 = 7$                       4)    $8 \div y = 4$

_____                        _____

5)    $5x \div 5 = 15$                      6)    $2(a \div 1) = 6$

_____                        _____

7)    $4y \div 2 = 16.6$                    8)    $16 \div 4x = 2$

_____                        _____

9)    $7(c \div 8) = 14$                   10)    $2(12 \div d) = 4$

_____                        _____


**Exercises 15**    Find the value of each variable.

1)    If $n = 3$, find the value of $18 \div n$.        2)    If $x = 12$, find the value of $x \div 3$.

_____                        _____

3)    If $x = 6$, find the value of $12 \div (0.5x)$.   4)    If $y = 36$, find the value of $y \div 4$.

_____                        _____

5)    If $x = 2$, find the value of $2(4 \div x)$.      6)    If $x = 3$, find the value of $24 \div x + 3$.

_____                        _____

7)    If $y = 9.6$, find the value of $6(y \div 1.2$.   8)    If $x = 5$, find the value of $28 \div x$.

_____                        _____

9)    If $2x = 3$, find the value of $12 \div (2x)$.   10)    If $3x = 4$, find the value of $(3x) \div 2$.

_____                        _____

**Exercises 16**    Solve each equation.

**1)**    $6 \div 2n = 6$                        **2)**    $36 \div 3x = 6$

_____                              _____

**3)**    $x \div 3 = 6.2$                      **4)**    $32.8 \div n = 4.1$

_____                              _____

**5)**    $2c \div 5 = 14$                      **6)**    $22.4 \div d = 8$

_____                              _____

**7)**    $5x \div 2.5 = 10$                    **8)**    $x \div 4.5 = 6$

_____                              _____

**9)**    $(y - 2) \div 5 = 5$                  **10)**    $4 \div \dfrac{x}{2} = 4$

_____                              _____


**Exercises 17**    Find the value of each variable.

**1)**  If $n = 3$, find the value of $18 \div (2n)$.     **2)**  If $2x = 12$, find the value of $2(x \div 3)$.

_____                      _____

**3)**  If $2x = 6$, find the value of $12 \div (4x)$.    **4)**  If $6y = 4$, find the value of $6y \div 2$.

_____                      _____

**5)**  If $x = 5$, find the value of $2(10 \div x)$.     **6)**  If $c = 4$, find the value of $28 \div c + 10$.

_____                      _____

**7)**  If $2y + 1 = 7$, find the value of $6(y \div 3)$.   **8)**  If $x \div 5 = 12$, find the value of $3(x \div 5)$.

_____                      _____

**9)**  If $x \div 2 = 8$, find the value of $x \div 2 - 1$.  **10)**  If $6 \div x = 4$, find the value of $10(6 \div x) + 4$.

_____                      _____

**SELF-TEST**

1. Which of the following expressions represents "the product of $\frac{1}{3}$ and $x$"?

   **A.** $\frac{1}{3} \div x$                              **B.** $\frac{1}{3} - x$

   **C.** $\frac{1}{3} + x$                              **D.** $\frac{1}{3} \times x$

2. Which of the following expressions represents "the product of $\frac{1}{2}N$ and 1"?

   **A.** $\frac{1}{2}N + 1$                              **B.** $N \div 1$

   **C.** $\frac{1}{2}N$                                       **D.** $N - 1$

3. What is the value of $x$ for the equation below?
$$\frac{1}{5}(3 \times x) = 9$$

   **A.** 10                                          **B.** 15
   **C.** 18                                          **D.** 20

4. What is the value of $x$ for the equation below?
$$8 \times x = 3\frac{1}{5}$$

   **A.** $\frac{1}{5}$                                        **B.** $\frac{2}{5}$

   **C.** $\frac{3}{5}$                                        **D.** $\frac{4}{5}$

5. What is the value of $x$ for the equation below?
$$9 \times \frac{1}{3}x = 3$$

   **A.** 1                                           **B.** 2

   **C.** 3                                           **D.** $\frac{1}{3}$

6. Marcus is dividing his marbles into 2 jars, which are marked JA and JB. JB has 5 more marbles than JA. If he has 18 marbles, which expression could be used to show

how many marbles are in each jar?

    **A.** $18 = (x - 5) + x$                             **B.** $18 = (x + 5) + x$

    **C.** $18 = (5x) + x$                                 **D.** $18 = (x + 5) + x$

7. At a grocery market, one bunch of bananas costs \$3.20. If Judy paid \$12.80, what is the equation could be used to find how many bunches she bought?

    **A.** $12.80 = 3.20x$                             **B.** $12.80 = 3.20 \div x$

    **C.** $12.80 = 3.20 + x$                           **D.** $12.80 + 3.20 = x$

8. J.R. is counting his quarters. He has organized his coins so that there are 6 quarters in one bag. If J.R. has \$9 in total, what equation could be used to find how many bags he has?

    **A.** $6 \times 0.25x = 9$                             **B.** $9 = 6x + 0.25$

    **C.** $6 + 0.25x = 9$                             **D.** $9 = 6x$

9. Two tables are set out for a dinner. On each table are two plates of egg roll. The first table has $\frac{1}{3}$ as many egg rolls than the second table. If there are 12 egg rolls in total, what is the equation used to find how many egg rolls there are on the second table?

    **A.** $12 = \frac{1}{3}(x)$                             **B.** $12 = \frac{1}{3}(x) \times x$

    **C.** $12 = \frac{1}{3}(x) - x$                          **D.** $12 = \frac{1}{3}(x) + x$

10. Which of the following expressions represents "$\frac{1}{5}$ divided by $x$"?

    **A.** $\frac{1}{5} \div x$                                **B.** $\frac{1}{5} - x$

    **C.** $\frac{1}{5} + x$                               **D.** $\frac{1}{5} \times x$

11. Which of the following expressions represents "$\frac{1}{2}y$ divided by 3"?

    **A.** $3 \div y$                                   **B.** $\frac{1}{2}y \div 3$

    **C.** $\frac{1}{2} + y$                              **D.** $\frac{1}{2}y \times 3$

12. What is the value of $x$ for the equation below?
$$\frac{1}{3}(9 \div x) = 1$$

    **A.** 1                                    **B.** 2
    **C.** 3                                    **D.** 4

13. What is the value of $x$ for the equation below?
$$\frac{1}{2}x \div 4 = 6$$

    **A.** 12                                  **B.** 24
    **C.** 48                                  **D.** 36

14. What is the value of $x$ for the equation below?
$$(9 \div x) = \frac{1}{3}$$

    **A.** 3                                    **B.** 4
    **C.** 5                                    **D.** 27

15. What is the value of $x$ for the equation below?
$$\frac{1}{2}x \div 2 = 2$$

    **A.** 2                                    **B.** 4
    **C.** 6                                    **D.** 8

16. If $3z = 5$, what is the value of $(13 - 6z) + 6$?

    **A.** 2                                    **B.** 3
    **C.** 9                                    **D.** 12

17. If $2x - 3 = 1$, what is the value of $2x - 3 + 2(2x - 3)$?

    **A.** 0                                    **B.** 1
    **C.** 2                                    **D.** 3

18. If $3 \div x = 2$, what is the value of $3 \div x + 2$?

    **A.** 2                                    **B.** 3
    **C.** 4                                    **D.** 5

19. If $3(x \div 2) = 9$, what is the value of $(x \div 2) - 1$?

        A. 0                                     B. 1
        C. 2                                     D. 3

\* For Exercises **20-21**. Joey has 21 toy cars he wants to put in several boxes. He puts 6 cars in each box and has 3 cars remaining.

20. Which of the following equations could be used to find that how many boxes he used?

        A. $21 \div x = 6 + 3$                     B. $21 \times 6 = x + 3$
        C. $(21 - 3) \times x = 6$                 D. $(21 - 3) \div x = 6$

21. How many boxes did he use?

        A. 2                                       B. 3
        C. 4                                     D. 5

22. At a school, 79 students and 7 teachers are going a field trip. A bus can seat 28 people. Which equation could be used to show how many buses are needed to take every student and teacher?

        A. $(79 + 7) \div x = 28r2$              B. $(79 + 7 - 2) \div x = 28$
        C. $(79 + 7) \div x = 28 + \dfrac{2}{x}$         D. All of the above.

23. How many buses will be needed to take every student and teacher?

        A. 2                                       B. 3
        C. 4                                     D. 5

24. If the quotient of an equation is 6, what is the value of the divisor given the dividend is 24?

        A. 3                                       B. 4
        C. 6                                   D. $\dfrac{1}{4}$

### 3. Using Two Variables

**2–11.**  Given $2x + y = 20$ and $x = 9$, what is the value of the $y$?

> **SOLUTION**
>
> a) There are two variables in the equation, but only $x$ is given. So, you can replace $x$ with 9 and then solve for $y$.
>
> $\quad\quad 2 \times 9 + y = 20$         Substitute $x$ with **9**.
> $\quad\quad 18 + y = 20$             Multiply 2 by 9.
>
> b) If you would like to solve for $y$ in the equation $18 + y = 20$, then first subtract 18 from both sides.
>
> $\quad\quad (\textbf{-18}) + 18 + y = 20 - \textbf{18}$     Subtract **18** from both sides.
> $\quad\quad \cancel{-18} + \cancel{18} + y = 20 - 18$     Cancel out -18 and 18.
> $\quad\quad y = 2$                   Simplify. $20 - 18 = 2$
> So, the value of $y$ is 2.

**Exercises 18**    Find the value of each unknown variable

1)  If $y = 3$, $x + 2y = 42$

2)  If $x = 4$, $2x + y = 115$

3)  If $y = 2$, $128 = 24y + x$

4)  If $a = 8$, $3a + 56 = b$

5)  If $n = 5$, $102 = 4n + 2m$

6)  If $x = 6$, $81 = 5y + x$

7)  If $y = 4$, $2x + 3y = 28$

8)  If $x = 4$, $2(2x + y) = 105$

9)  If $y = 2$, $56 = 4(2y + x)$

10) If $a = 3$, $3(a + 5) = b$

11) If $n = 5$, $38 = 2(n + 2m)$

12) If $x = 3$, $60 = 5(y + 2x)$

**Exercises 19**   Find the value of each expression.

1)  If $x = 2$ and $y = 3$, find the value of $5 \times x + 6 - y$.

2)  If $m = 1$ and $n = 7$, find the value of $2m - 1 + 2n$.

_____

3)  If $x = 5$ and $y = 3$, find the value of $5 \times (2x - y)$.

4)  If $m = 4$ and $n = 9$, find the value of $6 + mn$.

_____

5)  If $k = 4$ and $l = 6$, find the value of $(2k - 1) + (2l + 1)$.

6)  If $p = 2$ and $q = 3$, find the value of $6 - (6p) \div q$.

_____

7)  If $s = 10$ and $t = 2$, find the value of $s \div (4t) + 4$.

8)  If $u = 6$ and $v = 4$, find the value of $(u \div 2) + uv$.

_____

9)  If $x = 5$ and $y = 2$, find the value of $2 + 2x - 3y$.

10)  If $2c = 8$ and $3d = 6$, find the value of $d + 4c - 8$.

_____

**Exercises 20**   Find the value of the unknown variable in each equation.

1)  If $x = \frac{1}{2}$, $78 = 6x + 3y$

2)  If $4x = 3$, $117 = 3y + 12x$

_____

3)  If $5y + 1 = 21$, $2x + 5y = 46$

4)  If $2x = 9$, $2(2x + y) = 32$

_____

5)  If $3x = 1$, $9 = 3(2y - 6x)$

6)  If $\frac{1}{2}x = 3$, $3(2x - 5) = b$

_____

7)  If $2x = 12$, $102 = 2(xy + 2x)$

8)  If $x = 3$, $20 = 2(y - 2x)$

_____

**Exercises 21**    Find the value of the unknown variable in each equation.

**1)**    If $y = 3$, $x - 4y = 27$                **2)**    If $n = 4$, $m - 6n = 74$

_____                    _____

**3)**    If $y = 7$, $7y - x = 25$                **4)**    If $x = 8$, $5x - 25 = y$

_____                    _____

**5)**    If $b = 4$, $9b - 4a = 12$                **6)**    If $y = 3$, $x - 3y = 47$

_____                    _____

**7)**    If $y = 2$, $3(x - 4y) = 21$              **8)**    If $n = 3$, $2(m - 6n) = 28$

_____                    _____

**9)**    If $y = 5$, $2(7y - x) = 40$              **10)**   If $x = 5$, $3(5x - 20) = y$

_____                    _____


**Exercises 22**    Find the value of each equation.

**1)**    If $x = 2$ and $y = 1$, find the value of $[(16$              **2)**    If $m = 2$ and $n = 3$, find the value of $6$
$\div (2x)] - y$.                                                      $+ 2(mn - n)$.

_____                    _____

**3)**    If $k - 1 = 3$ and $2l + 1 = 4$, find the              **4)**    If $p + 3 = 2$ and $q - 1 = 3$, find the
value of $2(k - 1) + (2l + 1)$.                        value of $6(p + 3) - 2(q - 1)$.

_____                    _____

**5)**    If $st = 4$ and $t = 2$, find the value of $(st)$              **6)**    If $2uv = 6$ and $v = 3$, find the value of
$+ (2s)$.                                              $2v + uv - u$.

_____                    _____

**7)**    If $xy - 2 = 1$ and $2y = 8$, find the value              **8)**    If $3(pq + 1) = 12$ and $2p = 6$, find the
of $2xy - 4 + (x + y)$.                                value of $2q + 8p - 6(pq + 1)$.

_____                    _____

**SELF-TEST**

1. If $k = 6$, what is the value of $27 - \frac{2}{3}(k + 6)$?

    **A.** 23                     **B.** 35
    **C.** 32                     **D.** 19

2. If $z = 2$, what is the value of $(3 - z) + \frac{2}{3}$?

    **A.** $1\frac{1}{3}$                     **B.** $2\frac{2}{3}$

    **C.** $1\frac{2}{3}$                     **D.** $2\frac{1}{3}$

3. Given the quotient of 18 and $x$ is increased by 8, what is the value if $x = 9$?

    **A.** 10                     **B.** 8
    **C.** 6                       **D.** 4

4. Given the product of $\frac{2}{7}$ and $K$ decreased by 2, what is the value if $K = 14$?

    **A.** 0                       **B.** 1
    **C.** 2                       **D.** 3

5. Given the sum of $\frac{2}{5}$ and $N$ increased by 5, what is the value if $N = 2$?

    **A.** 12                     **B.** 11
    **C.** 10                     **D.** 9

6. Given the difference between $x$ and 3 divided by 2, what is the value if $x = 3\frac{2}{5}$?

    **A.** $\frac{1}{5}$                     **B.** $\frac{2}{5}$

    **C.** $\frac{3}{5}$                     **D.** $\frac{4}{5}$

7.  If $z = 7$, what is the value of $\frac{1}{2}(2z + 6)$?

        **A.** 10                             **B.** 8
        **C.** 6                              **D.** 4

8.  If $x = 3$, what is the value of $\frac{2}{3}(9 - 3x) \times 6$?

        **A.** 0                              **B.** 1
        **C.** 2                              **D.** 3

9.  If $k = 6$, what is the value of N in $19 - (k + 7) = N$?

        **A.** 5                              **B.** 6
        **C.** 7                              **D.** 8

10.  If $c = 6$, what is the value of N in $N - 3c = 7$?

        **A.** 23                             **B.** 35
        **C.** 25                             **D.** 19

11.  If $z = 15$, what is the value of $k$ in $19 + k + 2z = 83$?

        **A.** 34                             **B.** 35
        **C.** 36                             **D.** 33

12.  If $k = 2$, what is the value of N in $57 - (14k + 7) = N$?

        **A.** 20                             **B.** 26
        **C.** 24                             **D.** 22

13.  Add the product of 2 and $K$ to the product of 4 and $N$. Given that $K = 4$, and the value of the equation is 20, what is the value of N?

        **A.** 1                              **B.** 2
        **C.** 3                              **D.** 4

14. Multiply the sum of 5 and $N$ to the difference of 10 and $K$.  Given that $K = 8$, and the value of the equation is 14, what is the value of N?

    A. 1                                    B. 2
    C. 3                                    D. 4

15. Subtract the quotient of 36 and $K$ to the product of 4 and $N$.  Given that $K = 2$, and the value of the equation is 6, what is the value of N?

    A. 1                                    B. 2
    C. 3                                    D. 4

16. If $x = 3$ and $y = 4$, what is the value of $(12 - 3x) + xy$?

    A. 14                                   B. 15
    C. 16                                   D. 17

17. If $2k - 1 = 5$, what is the value of $17 - 2(2k - 1) = M$?

    A. 7                                    B. 8
    C. 9                                    D. 10

18. If $5 - 3x = 3$ and $y + 5 = 4$, what is the value of $3(5 - 3x) + y + 5$?

    A. 16                                   B. 15
    C. 14                                   D. 13

19. If $3xy = 1$ and $y = 2$, what is the value of $1 - 3xy + 2y$?

    A. 1                                    B. 2
    C. 3                                    D. 4

20. If $2x = 3$ and $3 - xy = 1$, what is the value of N in $6x - (xy + 3) = N$?

    A. 2                                    B. 3
    C. 4                                    D. 4

### 4.  Understanding Coordinate Grids

**2–12.**  What represents $(x, y)$ on a coordinate grid?

> a)  $(x, y)$ represents the coordinates of a point.
> b)  The $x$-axis is the horizontal line and the $y$-axis is the vertical line.

**2–13.**  Plot Q(–2, –4) on a coordinate grid.

**SOLUTION**

a) First, look at A(–2, –4).

$(-2, -4)$

↑      ↑

|    $y$-coordinate

$x$-coordinate

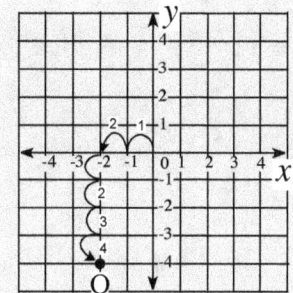

b)  i)   Start at point 0.
    ii)  Count 2 units to the left along $x$-axis.
    iii) Then count 4 units down $y$-axis.

**Exercises 23**    Plot the coordinates on a coordinate grid.

**1)** P(2, 4)

**2)** Q(–5, 2)

**3)** R(–2, –5)

**4)** S(2, –4)

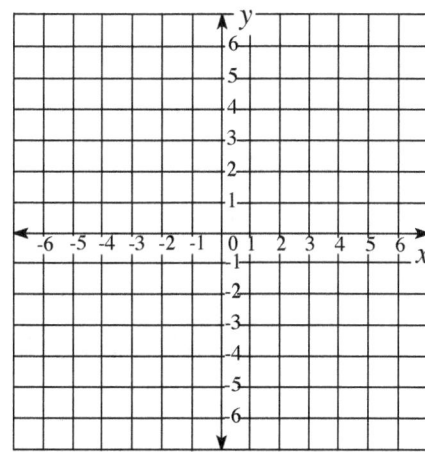

**Exercises 24**   Name the coordinates for each point.

**1)** H(    ,    )   **2)** E(    ,    )

**3)** X(    ,    )   **4)** A(    ,    )

**5)** G(    ,    )   **6)** O(    ,    )

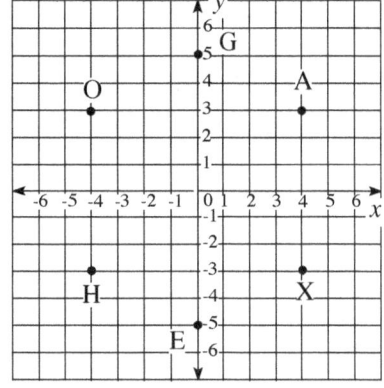

**Exercises 25**   Use the graph below in order to answer the following questions.

**1)** Which point is located at (−1, −3)?

**2)** What are the coordinates for R?

**3)** Which point is located at (−4, 2)?

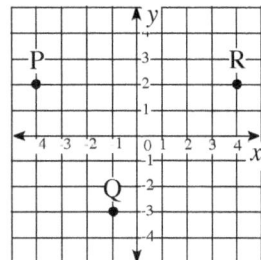

**Exercises 26**   Use the graph below to find the coordinates of the points.

**1)** Which point is located at (1, −3)?

**2)** What are the coordinates for D?

**3)** Which point is located at (−4, 4)?

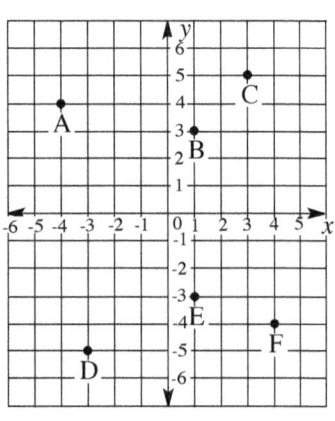

**Exercises 27**    Use the graph below in order to answer the following questions.

**1)** Draw A (–4, –4), B (–2, –2), and C (2, 2). Find the unknown *y*-coordinate of D (4, ?) so that D is on the same line as the other three points.

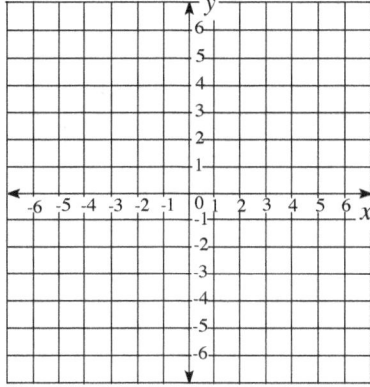

**2)** Draw A (-4, 5), B (–2, 4), and C (6, 0). Find the unknown *y*-coordinate of D (0, ?) so that D is on the same line as the other three points.

**Exercises 28**    Use the graph below in order to answer the following questions.

**1)** Draw A(–1, –3) and B(5, 6). Name the unknown *x*-coordinate of C(?, –6) so that C is on the same line as the other two points.

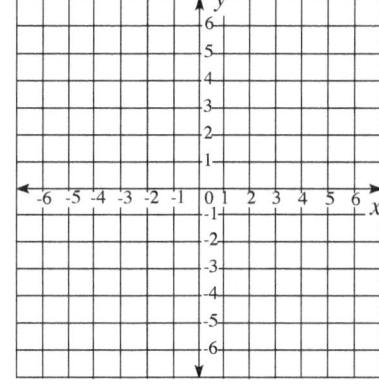

**2)** Draw A(–5, 5) and B(3, –1). Find the unknown *x* and *y* coordinates of C(?, ?) and D(?, ?) so that they are on the same line as A and B.

**2–14.** Find the distance between A and B on the coordinate grid.

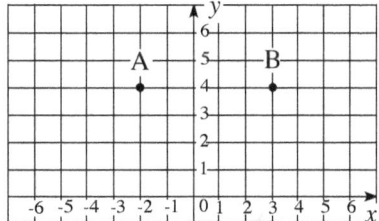

**SOLUTION**

You can use two ways to find the distance between A and B on the coordinate grid.

a) First, look at the coordinates of A and B.

   A(–2, 4) and B(3, 4)

i) When you look at the y-coordinates of A(–2, **4**) and B(3, **4**), they have the same y-coordinates.

ii) Therefore, only be concerned with the x-coordinates of A(**–2**, 4) and B(**3**, 4). Subtract the values of the x-coordinates between A and B or **3** to **–2**, which will be the distance. Write that $\overline{AB} = 3 - (-2) = 3 + 2 = 5$. Therefore, the distance between A(–2, 4) and B(3, 4) on the coordinate grid is 5.

b) Let's use a graph to count the units between the points.

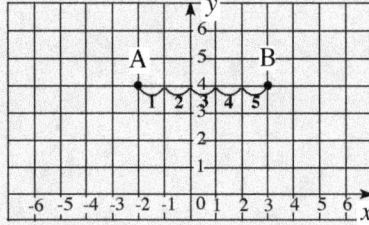

Therefore, the number of units between A(–2, 4) and B(3, 4) on the coordinate grid is 5.

**Exercises 29**    Use the graph below.

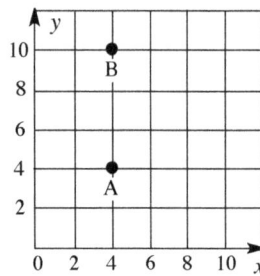

1) What are the coordinates for A and B?

2) What is the distance between A and B? Explain your answer.

**Exercises 30**    Use the graph below.

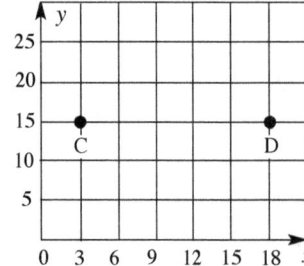

1) What are the coordinates for C and D?

2) What is the distance between C and D? Explain your answer.

**Exercises 31**   Use the graph below.

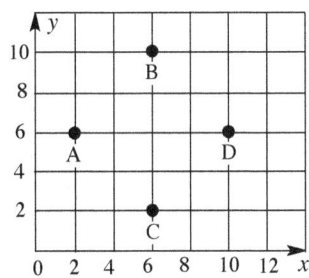

1) Name the point that is 8 units away from (6, 2).

2) Which point can be moved from (6, 10) to the point (10, 6)? Explain your answer.

**Exercises 32**   Find the distance between the points on the coordinate grid.

1)

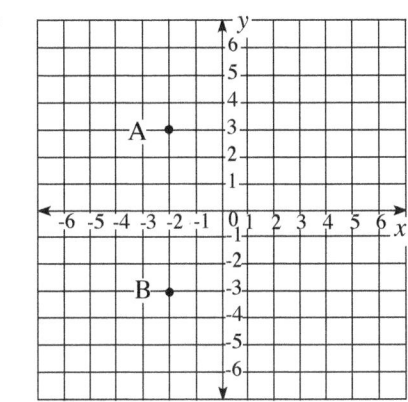

2)

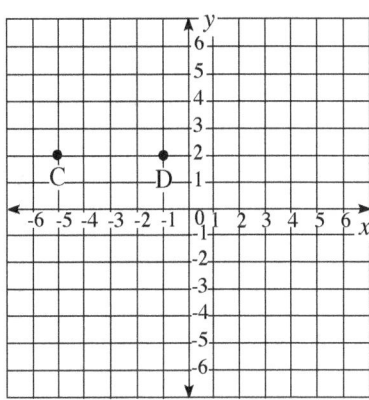

3)

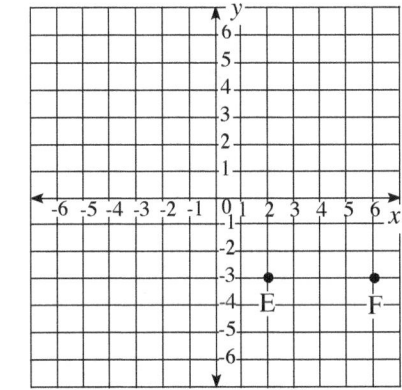

4)

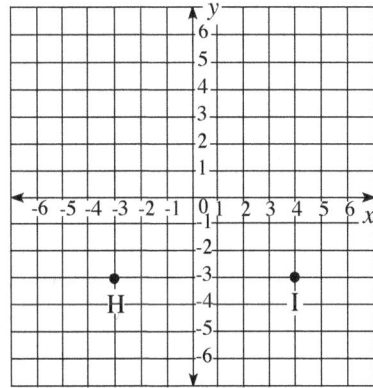

**Exercises 33**        Graph the coordinates and complete the function table.

|   | A | B | C | D |
|---|---|---|---|---|
| $x$ | 0 |   |   |   |
| $y$ |   |   |   |   |

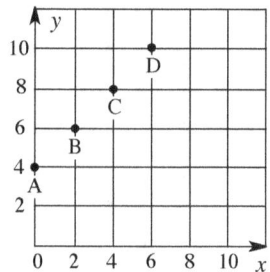

**Exercises 34**    Graph the coordinates and find the distance between each set of points.

**1)**  L(–5, 3) and M(2, 3)

**2)**  O(–5, –3) and L(–5, 3)

**3)**  O(–5, –3) and N(2, –3)

**4)**  N(2, –3) and M(2, 3)

**5)**  Find the perimeter of a figure.

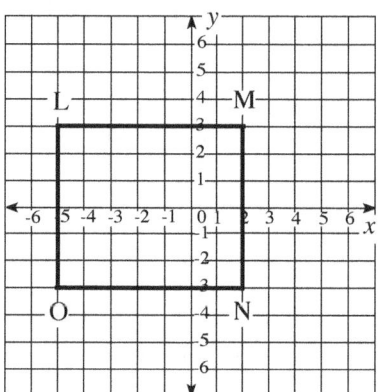

**Exercises 35**    Graph the coordinates and find the distance between each set of points.

**1)**  S(5, 3) and T(5, 0)

**2)**  U(–4, 3) and V(–4, 0)

**3)**  S(5, 3) and U(–4, 3)

**4)**  V(–4, 0) and T(5, 0)

**5)**  Name the shape of the figure after connecting each point.

**6)**  Find the perimeter of a figure.

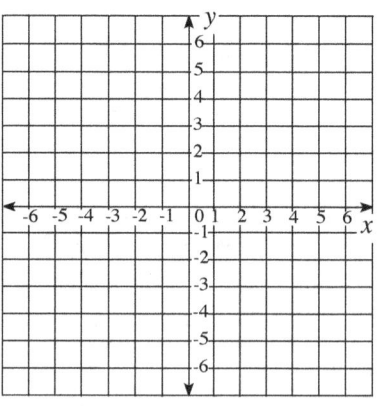

* For Exercises **1-4**, use the graph.

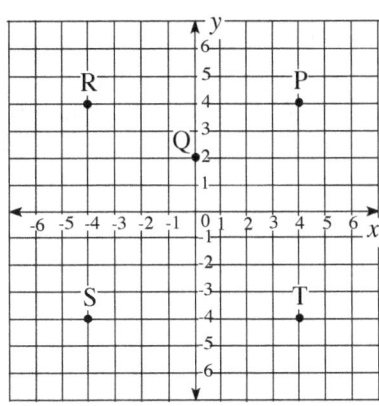

1.  Which of the following are the coordinates for R?

    **A.** (4, 4)          **B.** (4, −4)
    **C.** (−4, −4)      **D.** (−4, 4)

2.  Which of the following points is located at (4, −4)?

    **A.** P                **B.** R
    **C.** S                **D.** T

3.  Which of the following coordinates is at Q?

    **A.** (0, 2)                          **B.** (0, −2)
    **C.** (2, 2)                          **D.** (2, 0)

4.  Which of the following points is at (−4, −4)?

    **A.** Q                              **B.** R
    **C.** S                              **D.** T

* For Exercises **5-7**, use the graph to find each location in a town.

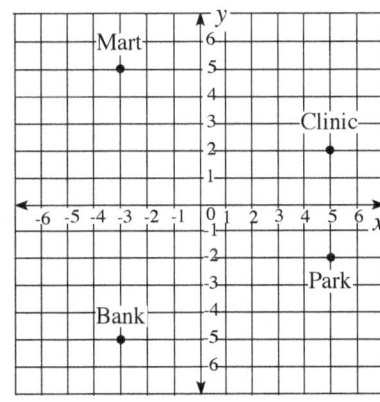

5.  What are the coordinates for the bank?

    **A.** (5, −3)          **B.** (−3, 5)
    **C.** (5, 3)          **D.** (−3, −5)

6.  What are the coordinates for the clinic?

    **A.** (5, −2)          **B.** (−2, 5)
    **C.** (5, 2)          **D.** (2, 5)

7.  Which of the following places is located at the point (–3, 5)?

        **A.** Mart                             **B.** Clinic

        **C.** Bank                             **D.** Park

\* For Exercises **8-10**, use the graph to find each location.

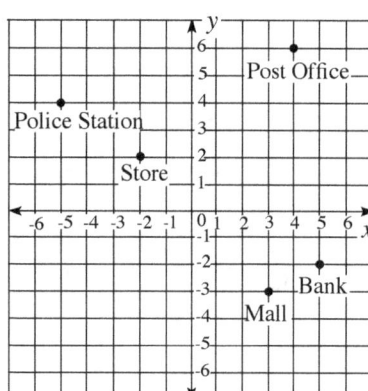

8.  What are the coordinates of the Post Office?

    **A.** (6, 4)         **B.** (4, 6)

    **C.** (–5, 4)      **D.** (–3, –3)

9.  Which of the following places is located at the point (5, –2)?

    **A.** Police Station

    **B.** Store

    **C.** Mall

    **D.** Bank

10.  What are the coordinates of the mall?

        **A.** (–1, 4)                         **B.** (3, –3)

        **C.** (–2, 4)                         **D.** (–3, 3)

\* For Exercises **11-14**, use the graph.

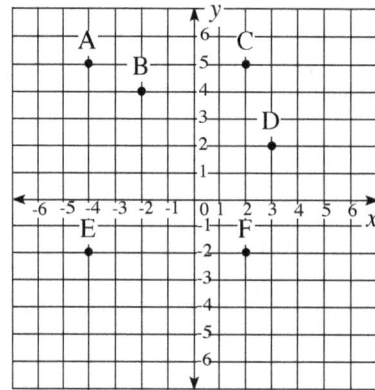

11.  Which of the following are the coordinates for A?

      **A.** (–4, 5)                    **B.** (–4, –5)
      **C.** (4, 5)                      **D.** (5, –4)

12.  Which of the following points is located at (2, 5)?

      **A.** A                        **B.** B
      **C.** C                        **D.** D

13.  What is the distance between A and C?

      **A.** 4 units                    **B.** 5 units
      **C.** 6 units                    **D.** 7 units

14.  What is the distance between C and F?

      **A.** 5 units                    **B.** 6 units
      **C.** 7 units                    **D.** 8 units

\*  For Exercises **15-22**, use the graph.

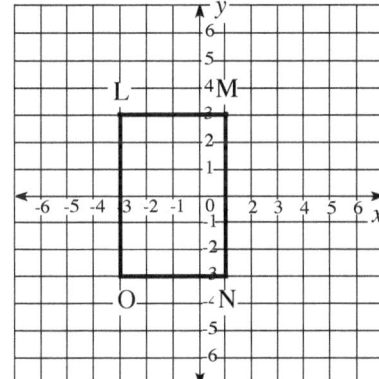

15.  Which of the following coordinates is at M?

      **A.** (–1, 3)                    **B.** (–3, –1)
      **C.** (3, 1)                      **D.** (1, 3)

16.  Which of the following points is located at (−3, −3)?

      **A.** L                              **B.** M
      **C.** O                              **D.** N

17.  What is the distance of $\overline{LM}$?

      **A.** 1 unit                         **B.** 4 units
      **C.** 6 units                      **D.** 10 units

18.  What is the distance of $\overline{MN}$?

      **A.** 1 unit                         **B.** 4 units
      **C.** 6 units                      **D.** 10 units

19.  What is the distance between (−3, 3) and (1, −3)?

      **A.** 1 unit                         **B.** 4 units
      **C.** 6 units                      **D.** 10 units

20.  Given two points located at (1, −3) and (1, 3), what is the name of the segment they form?

      **A.** $\overline{LM}$                         **B.** $\overline{NM}$
      **C.** $\overline{ON}$                      **D.** $\overline{OL}$

21.  Find the perimeter of LMON.

      **A.** 24 unit                        **B.** 20 units
      **C.** 18 units                    **D.** 12 units

22.  Find the area of LMON.

      **A.** 24 unit$^2$                      **B.** 20 units$^2$
      **C.** 18 units$^2$                  **D.** 12 units$^2$

## 5. Function Table

**2–15.** Use the given equation to make a function table.

$$y = 4 \times x$$

| x | 0 | 1 | 2 | 3 |
|---|---|---|---|---|
| y |   |   |   |   |

**SOLUTION**

a) Find the value for $y$.

$y = 4 \times x$ or $y = 4x$ (The sign ($\times$) can be omitted between the number and the variable).

| x | 0 | 1 | 2 | 3 |
|---|---|---|---|---|
| y |   |   |   |   |

$\Downarrow$ $y = 4 \times x$

| x | 0 | 1 | 2 | 3 |
|---|---|---|---|---|
| y | 0 | 4 | 8 | 12 |

i)  If $x = 0$, then $y = 0$.

    $y = 4 \times x$     Original equation

    $= 4 \times 0$     Substitute 0 for $x$.

    $= 0$

ii)  If $x = 1$, then $y = 4$.

    $y = 4 \times x$     Original equation

    $= 4 \times 1$     Substitute 1 for $x$.

    $= 4$

iii)  If $x = 2$, then $y = 8$.

    $y = 4 \times x$     Original equation

    $= 4 \times 2$     Substitute 2 for $x$.

    $= 8$

iv)  If $x = 3$, then $y = 12$.

    $y = 4 \times x$     Original equation

    $= 4 \times 3$     Substitute 3 for $x$.

    $= 12$

**2–16.** Use the given equation to make a function table.

$$y = 2x - 1$$

| x | 0 | 1 | 2 | 3 |
|---|---|---|---|---|
| y |   |   |   |   |

**SOLUTION**

Find the value for $y$.

$y = 2x - 1$ (The sign ($\times$) can be omitted between the number and the variable).

| x | 0 | 1 | 2 | 3 |
|---|---|---|---|---|
| y |   |   |   |   |

i)  If $x = 0$, then $y = -1$.

    $y = 2x - 1$     Original equation

    $= (2 \times 0) - 1$     Substitute 0 for $x$.

    $= -1$

ii)  If $x = 1$, then $y = 1$.

    $y = 2x - 1$     Original equation

$$y = 2x - 1$$

| $x$ | 0 | 1 | 2 | 3 |
|-----|---|---|---|---|
| $y$ | -1 | 1 | 3 | 5 |

$= (2 \times 1) - 1$    Substitute 1 for $x$.
$= 1$

iii) If $x = 2$, then $y = 3$.

$y = 2x - 1$        Original equation
$= (2 \times 2) - 1$    Substitute 2 for $x$.
$= 3$

iv) If $x = 3$, then $y = 5$.

$y = 2x - 1$        Original equation
$= (2 \times 3) - 1$    Substitute 3 for $x$.
$= 5$

**2–17.** Use the given equation to make a function table.

$$y = x \div 2$$

| $x$ | | | | |
|-----|---|---|---|---|
| $y$ | -2 | -1 | 0 | 1 |

**SOLUTION**

The function table is appeared $y$-value as an input number and looked $x$-value as an output number, but an equation shown like the other way. For your calculations, it can be rewritten.

$y = x \div 2$        Original equation

$y = \dfrac{1}{2}x$        Rewrite as a fraction.

$2y = \dfrac{1}{2}x \times 2$        Multiply each side by 2.

$2y = x$        Simplify 2 and 2 using the GCF of 2. $\dfrac{1}{2}x \times 2 = x$

So, you can use this equation to find the function table or you can use another method as shown in the following.

| $x$ | | | | |
|-----|---|---|---|---|
| $y$ | -2 | -1 | 0 | 1 |

$$2y = x$$

| $x$ | -4 | -2 | 0 | 2 |
|-----|---|---|---|---|
| $y$ | -2 | -1 | 0 | 1 |

i) If $y = -2$, then $x = -4$.

$x = 2y$        Original equation
$= 2 \times (-2)$    Substitute -2 for $y$.
$= -4$

ii) If $y = -1$, then $x = -2$.

$x = 2y$        Original equation
$= (2 \times -1)$    Substitute -1 for $y$.
$= -2$

iii) If $y = 0$, then $x = 0$.

$x = 2y$        Original equation
$= (2 \times 0)$    Substitute 0 for $y$.
$= 0$

iv) If $y = 1$, then $x = 2$.

$x = 2y$        Original equation
$= (2 \times 1)$    Substitute 1 for $y$.
$= 2$

**Exercises 36**    Use each given equation to make a function table.

**1)** $y = x$

| $x$ | −1 | 0 | 1 | 2 |
|---|---|---|---|---|
| $y$ | | | | |

**2)** $y = -2x$

| $x$ | 0 | 1 | 2 | 3 |
|---|---|---|---|---|
| $y$ | | | | |

**3)** $y = -x + 1$

| $x$ | −3 | −1 | 1 | 3 |
|---|---|---|---|---|
| $y$ | | | | |

**4)** $y = 2x - 1$

| $x$ | 2 | 4 | 6 | 8 |
|---|---|---|---|---|
| $y$ | | | | |

**5)** $y = -2x - 1$

| $x$ | | | | |
|---|---|---|---|---|
| $y$ | 1 | 3 | 5 | 7 |

**6)** $y = 2x + 1$

| $x$ | | | | |
|---|---|---|---|---|
| $y$ | −3 | −1 | 1 | 3 |

**Exercises 37**    Use each given equation to make a function table.

**1)** $y = 3x + 2$

| $x$ | | | | |
|---|---|---|---|---|
| $y$ | | | | |

**2)** $y = 4x - 1$

| $x$ | | | | |
|---|---|---|---|---|
| $y$ | | | | |

**3)** $y = 2x + 1$

| $x$ | −1 | 0 | 1 | 2 |
|---|---|---|---|---|
| $y$ | | | | |

**4)** $y = x - 3$

| $x$ | 0 | 1 | 2 | 3 |
|---|---|---|---|---|
| $y$ | | | | |

**5)** $y = -x + 1$

| $x$ | | | | |
|---|---|---|---|---|
| $y$ | −2 | −1 | 0 | 5 |

**6)** $3y = x$

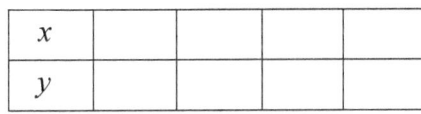

| $x$ | | | | |
|---|---|---|---|---|
| $y$ | −3 | −1 | 1 | 9 |

**Exercises 38**    Use each given equation to make a function table.

**1)** $4y = -x$

| $x$ | | | | |
|---|---|---|---|---|
| $y$ | -2 | -1 | 0 | 5 |

**2)** $5y = x$

| $x$ | | | | |
|---|---|---|---|---|
| $y$ | -2 | -1 | 0 | 5 |

**3)** $2y = x$

| $x$ | | | | |
|---|---|---|---|---|
| $y$ | -3 | -1 | 1 | 9 |

**4)** $2(1 + y) = x$

| $x$ | | | | |
|---|---|---|---|---|
| $y$ | -2 | -1 | 0 | 5 |

**5)** $y = \frac{1}{3}x + 1$

| $x$ | | | | |
|---|---|---|---|---|
| $y$ | -2 | -1 | 0 | 5 |

**6)** $y = -x + 1$

| $x$ | | | | |
|---|---|---|---|---|
| $y$ | -2 | -1 | 0 | 5 |

**Exercises 39**    Use each given equation to make a function table.

**1)** $y = -x + 3$

| $x$ | | | | |
|---|---|---|---|---|
| $y$ | -2 | -1 | 0 | 5 |

**2)** $y = x + 3$

| $x$ | | | | |
|---|---|---|---|---|
| $y$ | -2 | -1 | 0 | 5 |

**3)** $2y + 3 = -x$

| $x$ | | | | |
|---|---|---|---|---|
| $y$ | -2 | -1 | 0 | 5 |

**4)** $2(1 + y) = x$

| $x$ | | | | |
|---|---|---|---|---|
| $y$ | -3 | -1 | 1 | 9 |

**5)** $2(y - 3) = x$

| $x$ | | | | |
|---|---|---|---|---|
| $y$ | -2 | -1 | 0 | 5 |

**6)** $y = x - 2$

| $x$ | | | | |
|---|---|---|---|---|
| $y$ | -3 | -1 | 1 | 9 |

1.  What is the unknown value in the function table below?

| $x$ | 3 | 4 | 5 | 6 |
|---|---|---|---|---|
| $y$ | 10 | | 16 | 19 |

    **A.** 11                                    **B.** 12
    **C.** 13                                    **D.** 14

2.  What is the unknown value in the function table below?

| $x$ | 7 | 9 | 11 | 12 |
|---|---|---|---|---|
| $y$ | 20 | | 32 | 35 |

    **A.** 22                                    **B.** 25
    **C.** 26                                    **D.** 29

3.  Use the function table below. If the value of $x$ is 15, what is the value of $y$?

| $x$ | 0 | 3 | 4 |
|---|---|---|---|
| $y$ | 4 | 10 | 12 |

    **A.** 24                                    **B.** 26
    **C.** 30                                    **D.** 34

4.  Use the function table below. If the value of $x$ is 2, what is the value of $y$?

| $x$ | 5 | 7 | 9 |
|---|---|---|---|
| $y$ | 6 | 10 | 14 |

    **A.** 0                                     **B.** 1
    **C.** 2                                     **D.** 4

5.  Which of the following function tables represent the equation $y = x + 3$?

**A.**

| $x$ | -3 | 0 | 3 |
|---|---|---|---|
| $y$ | -3 | 0 | 3 |

**B.**

| $x$ | 2 | 3 | 4 |
|---|---|---|---|
| $y$ | 6 | 7 | 8 |

**C.**

| $x$ | -1 | 0 | 2 |
|---|---|---|---|
| $y$ | 2 | 3 | 5 |

**D.**

| $x$ | -1 | 1 | 3 |
|---|---|---|---|
| $y$ | 0 | 2 | 5 |

6.  What is the unknown value in the function table below?

| $x$ | -3 | | 3 | 12 |
|---|---|---|---|---|
| $y$ | -1 | 0 | 1 | 4 |

A. -2                                             B. -1
C. 0                                              D. 1

7.  What is the unknown value in the function table below?

| $x$ | 0 | 3 | 6 | 12 |
|---|---|---|---|---|
| $y$ | -1 | 0 | 1 | |

A. 2                                              B. 3
C. 4                                              D. 5

8.  Use the function table below. Given the value of $y$ is 6, what is the value of $x$?

| $x$ | 2 | 4 | 6 |
|---|---|---|---|
| $y$ | 2 | 3 | 4 |

A. 7                                              B. 8
C. 9                                              D. 10

9.  Use the function table below. If the value of $y$ is 0, what is the value of $x$?

| $x$ | 0 | 2 | 4 |
|---|---|---|---|
| $y$ | -1 | 3 | 7 |

**A.** $\frac{1}{2}$                                                                  **B.** 1

**C.** 2                                                                              **D.** 3

10.  Which equation represents the function table below?

| $x$ | 0 | 1 | 2 |
|---|---|---|---|
| $y$ | 1 | 3 | 5 |

**A.** $y = 2x$                                             **B.** $y = 2x + 1$

**C.** $x = 2y + 1$                                         **D.** $x = 3 \times y - 1$

11.  Which equation represents the function table below?

| $x$ | 8 | 10 | 12 |
|---|---|---|---|
| $y$ | 5 | 6 | 7 |

**A.** $2(y - 1) = x$                                        **B.** $y = \frac{1}{2}x - 1$

**C.** $2(y + 1) = x$                                        **D.** $y = 2 \times x - 2$

12.  Which of the following function tables represent the equation $y = \frac{1}{4}x$?

**A.**

| $x$ | −4 | 0 | 4 |
|---|---|---|---|
| $y$ | −4 | 0 | 4 |

**B.**

| $x$ | 8 | 12 | 16 |
|---|---|---|---|
| $y$ | 2 | 3 | 4 |

**C.**

| $x$ | −4 | 0 | 4 |
|---|---|---|---|
| $y$ | −1 | 1 | 2 |

**D.**

| $x$ | −4 | 4 | 8 |
|---|---|---|---|
| $y$ | −1 | 0 | 2 |

13.  Which of the following function tables represent the equation $y = -(2x - 1)$?

**A.**

| $x$ | −1 | 0 | 1 |
|---|---|---|---|
| $y$ | −3 | 0 | 3 |

**B.**

| $x$ | 0 | 1 | 2 |
|---|---|---|---|
| $y$ | 5 | 6 | 7 |

**C.**

| $x$ | −3 | 0 | 3 |
|---|---|---|---|
| $y$ | 7 | −1 | 5 |

**D.**

| $x$ | −2 | −1 | 0 |
|---|---|---|---|
| $y$ | −3 | −2 | −1 |

## 6.  Making Equations

**2–18.**  Find the equation that is represented in the function table below.

| $x$ | 0 | 1 | 2 | 3 |
|---|---|---|---|---|
| $y$ | 0 | 3 | 6 | 9 |

SOLUTION

a) First, look at $x = 0$ in the function table.
   When $x = 0$, then $y = \mathbf{0}$.
   So, you can write the equation that is $y =$
   $? \, x + \mathbf{0}$ or $y = ? \, x$.

| $x$ | **0** | 1 | 2 | 3 |
|---|---|---|---|---|
| $y$ | **0** | 3 | 6 | 9 |

b) Next, find the intervals between the
   values.
   Check the relationship between the
   values of $x$ and $y$.

The values of $x$ constantly increase by 1.
The values of $y$ constantly increase by 3.
Now, you can find the equation, which is
$y = \mathbf{3}x + 0$ or $y = \mathbf{3}x$.

| $x$ | 0 | 1 | 2 | 3 |
|---|---|---|---|---|
| $y$ | 0 | 3 | 6 | 9 |

**2–19.**  Find the equation that is represented in the function table below.

| $x$ | 0 | 1 | 2 | 3 |
|---|---|---|---|---|
| $y$ | −2 | 1 | 4 | 7 |

SOLUTION

a) First, look at $x = 0$ in the function table.
   When $x = 0$, then $y = \mathbf{-2}$.
   So, you can write the equation that is $y = ?$
   $x - \mathbf{2}$.

| $x$ | **0** | 1 | 2 | 3 |
|---|---|---|---|---|
| $y$ | **−2** | 1 | 4 | 7 |

b) Next, find the intervals between the
   values.
   Check the relationship between the
   values of $x$ and $y$.

The values of $x$ constantly increase by 1.
The values of $y$ constantly increase by 3.
Now, you can find the equation, which is
$y = \mathbf{3}x - 2$.

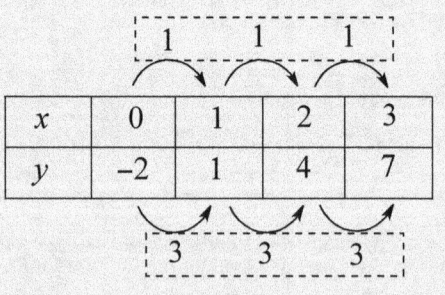

| $x$ | 0 | 1 | 2 | 3 |
|---|---|---|---|---|
| $y$ | −2 | 1 | 4 | 7 |

**2–20.** Find the equation of the function table.

| $x$ | −3 | 0 | 3 | 6 |
|---|---|---|---|---|
| $y$ | −1 | 0 | 1 | 2 |

**SOLUTION**

a) First, look at $x = 0$ in the function table.
   When $x = 0$, then $y = \mathbf{0}$. So, you can write
   the equation that is $y = ?x + \mathbf{0}$ or $y = ?x$.

| $x$ | −3 | **0** | 3 | 6 |
|---|---|---|---|---|
| $y$ | −1 | **0** | 1 | 2 |

b) Next, find the intervals between the values.
   Check the relationship between the values of $x$ and $y$.

The values of $x$ constantly increase by **3**.
The values of $y$ constantly increase by **1**.
Now, you can find the equation, which is **3**$y$ $= x$ or $y = \dfrac{1}{3}x$.

3  3  3

| $x$ | −3 | 0 | 3 | 6 |
|---|---|---|---|---|
| $y$ | −1 | 0 | 1 | 2 |

1  1  1

**2–21.** Find the equation of the function table.

| $x$ | −4 | −2 | 0 | 2 |
|---|---|---|---|---|
| $y$ | −1 | 0 | 1 | 2 |

**SOLUTION**

a) First, look at $x = 0$ in the function table.
   When $x = 0$, then $y = \mathbf{1}$. So, you can write
   the equation that is $y = ?x + \mathbf{1}$ or $y - 1 = x$.

| $x$ | −4 | −2 | **0** | 2 |
|---|---|---|---|---|
| $y$ | −1 | 0 | **1** | 2 |

b) Next, find the intervals between the values.

   Check the relationship between the values of $x$ and $y$.

The values of $x$ constantly increase by **2**.
The values of $y$ constantly increase by **1**.
Now, you can find the equation, which is **2**($y -$ $1) = x$ or $y = \dfrac{1}{2}x + 1$.

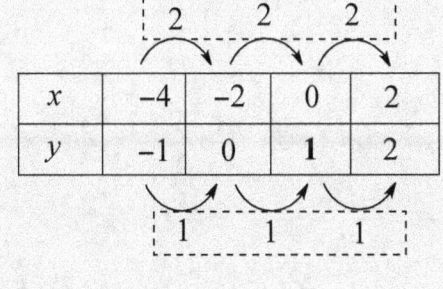

2  2  2

| $x$ | −4 | −2 | 0 | 2 |
|---|---|---|---|---|
| $y$ | −1 | 0 | 1 | 2 |

1  1  1

**SELF-TEST**

1.  What is the value of (Δ) given the information from the table below?

| $x$ | 0 | 1 | 2 | 3 |
|---|---|---|---|---|
| $y$ | 0 | 2 | 4 | 6 |

$y = (Δ)x$

A. 1                                     B. 2
C. 3                                     D. 4

2.  What is the value of (Δ) given the information from the table below?

| $x$ | −2 | −1 | 0 | 1 |
|---|---|---|---|---|
| $y$ | −4 | −2 | 0 | 2 |

$y = (Δ)x$

A. 1                                     B. 2
C. 3                                     D. 4

3.  What are the values of (Δ) and ('?) given the information from the table below?

| $x$ | −1 | 0 | 1 | 2 |
|---|---|---|---|---|
| $y$ | −3 | −1 | 1 | 3 |

$y = (Δ)x − (?)$

A. Δ = −2, ? = 1                         B. Δ = 3, ? = −1
C. Δ = 2, ? = 1                          D. Δ = 3, ? = 1

4.  What are the values of (Δ) and (?) given the information from the table below?

| $x$ | 0 | 1 | 2 | 3 |
|---|---|---|---|---|
| $y$ | 2 | 3 | 4 | 5 |

$y = (Δ)x + (?)$

A. Δ = 1, ? = 2                          B. Δ = 2, ? = 2
C. Δ = −1, ? = 2                         D. Δ = 2, ? = −2

5.  What is the value of (Δ) given the information from the table below?

| $x$ | $-1$ | 0 | 1 | 2 |
|---|---|---|---|---|
| $y$ | $-6$ | 0 | 6 | 12 |

$y = (\Delta)x$

**A.** 3                                    **B.** 4
**C.** 5                                    **D.** 6

6.  What is the equation that represents the following information from the table below?

| $x$ | $-1$ | 0 | 1 | 2 |
|---|---|---|---|---|
| $y$ | $-2$ | 1 | 4 | 7 |

**A.** $y = 3x + 1$                              **B.** $y = x - 1$
**C.** $y = 6x + 1$                              **D.** $y = 2x$

7.  What is the equation that represents the following information from the table below?

| $x$ | $-2$ | $-1$ | 0 | 1 |
|---|---|---|---|---|
| $y$ | $-4$ | $-3$ | $-2$ | $-1$ |

**A.** $y = 2x + 1$                              **B.** $y = x - 2$
**C.** $y = x + 4$                               **D.** $y = 2x$

8.  Which equation represents the function table below?

| $x$ | 0 | 1 | 2 | 3 |
|---|---|---|---|---|
| $y$ | 0 | 4 | 8 | 12 |

**A.** $y = 3x + 1$                              **B.** $y = 5x - 1$
**C.** $y = x + 3$                               **D.** $y = 4x$

9.  Which equation represents the function table below?

| $x$ | 0 | 2 | 4 | 6 |
|---|---|---|---|---|
| $y$ | 7 | 9 | 11 | 13 |

**A.** $y = x + 1$                               **B.** $y = x + 7$
**C.** $y = 6x + 1$                              **D.** $y = 7x + 1$

10. Which equation represents the function table below?

| $x$ | 0 | 1 | 2 | 3 |
|---|---|---|---|---|
| $y$ | −2 | −1 | 0 | 1 |

    **A.** $y = 2x - 2$                **B.** $y = -1x - 1$
    **C.** $y = x - 2$                 **D.** $y = 3x - 2$

11. Which equation represents the function table below?

| $x$ | 0 | 1 | 2 | 3 |
|---|---|---|---|---|
| $y$ | 1 | 7 | 13 | 19 |

    **A.** $y = x + 1$                **B.** $y = x + 7$
    **C.** $y = 6x + 1$               **D.** $y = 7x + 1$

12. Which equation represents the function table below?

| $x$ | −1 | 1 | 3 |
|---|---|---|---|
| $y$ | −2 | 4 | 10 |

    **A.** $y = 2x + 1$               **B.** $y = x - 1$
    **C.** $y = x + 4$                **D.** $y = 3x + 1$

13. Which equation represents the function table below?

| $x$ | −1 | 1 | 3 |
|---|---|---|---|
| $y$ | −5 | −3 | −1 |

    **A.** $y = -2x - 4$            **B.** $y = x - 4$
    **C.** $y = 2x - 4$              **D.** $y = 2x - 3$

14. Which equation represents the function table below?

| $x$ | −1 | 2 | 3 | 5 |
|---|---|---|---|---|
| $y$ | −1 | 8 | 11 | 17 |

**A.** $y = 3x + 2$  
**C.** $y = 9x + 2$

**B.** $y = 2x - 2$  
**D.** $y = 4x$

15. Which equation represents the function table below?

| $x$ | −1 | 0 | 1 | 2 |
|---|---|---|---|---|
| $y$ | −2 | −1 | 0 | 1 |

**A.** $y = x - 1$  
**C.** $y = 2x - 1$

**B.** $y = x + 1$  
**D.** $y = 2x + 1$

16. What are the values of $x$ and $y$ so that $y = x + 3$?

**A.**

| $x$ | −3 | 0 | 3 |
|---|---|---|---|
| $y$ | −3 | 0 | 3 |

**B.**

| $x$ | 2 | 3 | 4 |
|---|---|---|---|
| $y$ | 6 | 7 | 8 |

**C.**

| $x$ | −1 | 0 | 1 |
|---|---|---|---|
| $y$ | 2 | 3 | 4 |

**D.**

| $x$ | −1 | 1 | 3 |
|---|---|---|---|
| $y$ | 0 | 2 | 5 |

17. Which equation represents the function table below?

| $x$ | −2 | −1 | 0 | 1 |
|---|---|---|---|---|
| $y$ | −7 | −5 | −3 | −1 |

**A.** $y = 2x + 1$  
**C.** $y = 2x - 3$

**B.** $y = 2x - 2$  
**D.** $y = 4x$

18. Which equation represents the function table below?

| $x$ | 4 | 5 | 7 | 9 |
|---|---|---|---|---|
| $y$ | 11 | 14 | 20 | 26 |

**A.** $y = 3x + 1$  
**C.** $y = 2x + 3$

**B.** $y = 3x - 1$  
**D.** $y = 2x + 4$

# CHAPTER 3
# Decimals and Fractions

In this chapter, you will solve problems involving adding, subtracting, multiplying, and dividing decimals and fractions.

## 1. Place Value
### 3–1. Number Sense: Operation with Decimals

a) Write the whole number part located left of the decimal point
b) Write "**and**" to represent the decimal point.
c) Write the decimal number located right of the expression.

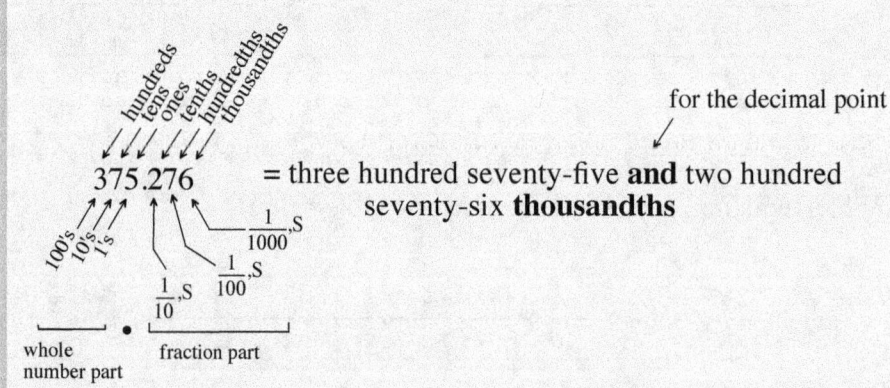

375.276 = three hundred seventy-five **and** two hundred seventy-six **thousandths**

for the decimal point

Expanded Form: $300 + 70 + 5 + \dfrac{2}{10} + \dfrac{7}{100} + \dfrac{6}{1000}$

**Exercises 1**    Write each decimal in its word form.

**1)** $0.05 =$

_____

Expanded Form:

_____

**2)** $2.06 =$

_____

Expanded Form:

_____

**3)**  83.08  =

_____

Expanded Form:

_____

**4)**  243.429 =

_____

Expanded Form:

_____

**5)**  0.70 =

_____

Expanded Form:

_____

**6)**  0.707 =

_____

Expanded Form :

_____

**Exercises 2**   Write each number in its standard form.

**1)**  six hundred nine and five hundred one thousandths

_____

**2)**  six thousandths

_____

**3)**  five and twenty-six hundredths

_____

**4)**  two hundred and one tenth

_____

## 2.  Estimating Sums and Differences

**3-2.**  Estimate the sum of 18.74 + 23.49.

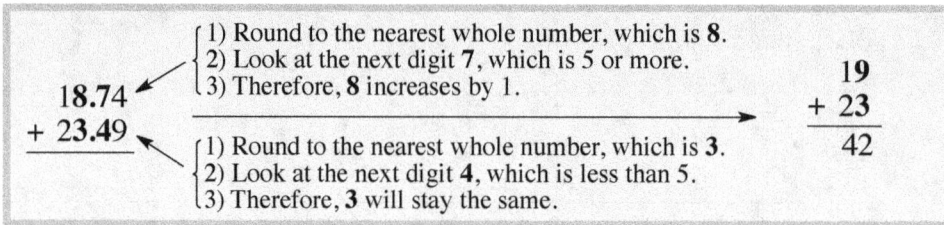

**Exercises 3**    Estimate the sum or difference of each expression.

**1)**    $3.75 – $1.20

**2)**    62.497 – 9.039

**3)**    632.93 + 75.39

**4)**    $38.46 + $84.82

**5)**    4.4 + 11.7

**6)**    63.52 + 27.49

**7)**    $19.59 + $53.85

**8)**    $729.49 – $285.79

**9)**    48.46 + 19.27

**10)**    14.54 + 9.47

**11)**    1.29 + 0.71

**12)**    34.62 + 29.29

## 3. Comparing Decimals

**3-3.** Comparing decimals

> a. First, compare the tenth place of both numbers and determine if they are different.
> b. Second, if they are the same number, then compare the hundredth digits with each other. If they are identical, then continue until you find a place value where the digits are different.
> c. Once you found the different numbers, compare them to determine the greatest and least of the decimals.

**3-4.** Compare each pair of decimals.

$$8.025 \ \square \ 8.052$$

**SOLUTION**

| $8 = 8$ | compare to next digit $\Rightarrow$ | $0.0 = 0.0$ | compare to next digit $\Rightarrow$ | $0.02 < 0.05$ |
|---|---|---|---|---|
| $8.025 \ \square \ 8.052$ | | $8.025 \ \square \ 8.052$ | | $8.025 \ \boxed{<} \ 8.052$ |

Now, you can decide which decimal number is greater. So, 8.025 is less than 8.052 or 8.052 is greater than 8.025.

$$8.025 \ \boxed{<} \ 8.052$$

**Exercises 4**    Compare the decimals with each other.

**1)**   2.708 _____ 2.709

**2)**   0.72 _____ 0.721

**3)**   $14.082 \ \underline{\hspace{1cm}} \ 14\dfrac{82}{100}$

**4)**   $84.19 \ \underline{\hspace{1cm}} \ 84\dfrac{19}{100}$

**5)**   $0.25 \ \underline{\hspace{1cm}} \ \dfrac{1}{4}$

**6)**   2% _____ 0.03

**7)**   2.708 _____ 2.709

**8)**   $0.76 \ \underline{\hspace{1cm}} \ \dfrac{3}{4}$

**9)**   $2\dfrac{2}{100} \ \underline{\hspace{1cm}} \ 2.20$

**10)**   22% _____ 2.22

**11)**   $1.500 \ \underline{\hspace{1cm}} \ \dfrac{8}{5}$

**12)**   1.9% _____ 0.019

**Exercises 5**    List the values from greatest to least.

1)   $2\frac{1}{10}$, 2.05, 2.11, $2\frac{1}{8}$

2)   10.4, 0.95, −10.99, −5.09

_____

_____

3)   $2\frac{1}{4}$, $2\frac{1}{3}$, 2.15, 2.17, 2.27

4)   0.359,  0.38,  0.392,  0.387, $\frac{8}{5}$

_____

_____

5)   92.648,  92.6489,  $91\frac{4}{5}$,  91.998

6)   2.3, 2.35, $2\frac{36}{100}$, $\frac{36}{10}$, 2.05

_____

_____

**Exercises 6**    Use the number lines to identify each unknown value.

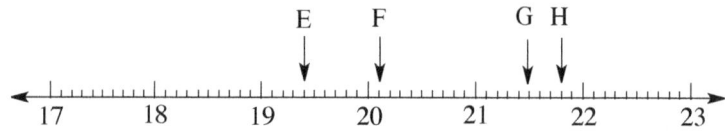

1)  E = _____

2)  F = _____

3)  G = _____

4)  H = _____

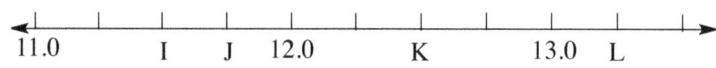

5)  I = _____

6)  J = _____

7)  K = _____

8)  L = _____

**SELF-TEST**

1.  Which of the following is the correct word form for the decimal below?
    37.92

    **A.**  thirty-seven and ninety-two tenths
    **B.**  thirty-seven and ninety-two hundredths
    **C.**  thirty-seven, ninety-two hundredths
    **D.**  thirty-seven, ninety-two tenths

2.  Which of the following is the correct expanded form for the decimal below?
    201.108

    **A.**  $200 + 1 + \dfrac{1}{10} + \dfrac{8}{100}$          **B.**  $201 + \dfrac{1}{10} + \dfrac{0}{100} + \dfrac{8}{1000}$

    **C.**  $200 + 1 + \dfrac{1}{10} + \dfrac{8}{1000}$          **D.**  $200 + 10 + \dfrac{1}{10} + \dfrac{8}{1000}$

3.  Which of the following is the correct standard form for the expanded form below?
    $$900 + 60 + \dfrac{7}{100} + \dfrac{1}{1000}$$

    **A.**  960.701                    **B.**  96.701
    **C.**  960.71                     **D.**  960.071

4.  Which of the following is the correct standard form for the number below?
    four and eleven thousandths

    **A.**  4.101                      **B.**  41.01
    **C.**  4.110                      **D.**  4.011

5.  Which of the following is the correct standard form for the number below?
    One hundred two and five thousandths

    **A.**  102.500                    **B.**  102.05
    **C.**  102.005                    **D.**  120.050

6.  What is the best estimated difference for the expression below?
    26.93 – 17.49

    **A.**  9.44                       **B.**  11.00
    **C.**  10.40                      **D.**  10

7.  Branny deposited $15.25 in his bank account on Saturday, $23.90 on Monday, and $11.50 on Tuesday. Which of the following is the best estimate for the total sum of the money she deposited?

   A.  $50.00                    B.  $51.00
   C.  $52.00                    D.  $49.00

8.  Which of the following values is the best estimated sum of 84.79 + 34.48?

   A.  119.27                    B.  110
   C.  120                       D.  119

9.  Randy put 0.74 pounds of birdseed in the bird feeder. Two hours later, he found 0.49 pounds left in the feeder.  How many pounds of birdseed were eaten? Round to the nearest tenth.

   A.  0.2                       B.  0.3
   C.  0.4                       D.  0.5

10.  What is the best estimated sum of 2.58 + 25?

   A.  25                        B.  27
   C.  27.58                     D.  28

11.  Which is the best estimated sum of 18.79 – 8?

   A.  10.79                     B.  11
   C.  12                        D.  10

12.  Lily has $45.35 in her purse and paid $12.65 for a movie ticket. Which of the following is the best estimate for the amount of money she has now?

   A.  $32.70                    B.  $32.07
   C.  $32.00                    D.  $32.35

13.  Which of the following is incorrect?

   A.  5 < 5.000                 B.  202.370 > 202.037
   C.  9.27 is less than 9.272.  D.  0.2 is greater than 0.02

* Use the number line for Exercises **14-16**.

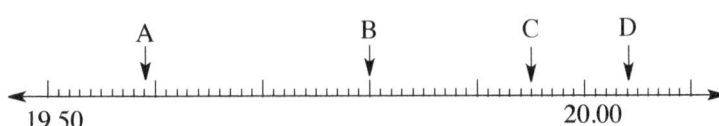

14. What is the relative position of A?

    **A.** 2.50                    **B.** 3.50
    **C.** 19.59                   **D.** 5.50

15. Name the relative position of 19.95.

    **A.** A                       **B.** B
    **C.** C                       **D.** D

16. Which of the following letters are 20.00 rounded to the nearest hundredths?

    **A.** A and B                 **B.** B and C
    **C.** C and D                 **D.** B, C, and D

* Use the number line for Exercises **17-19**.

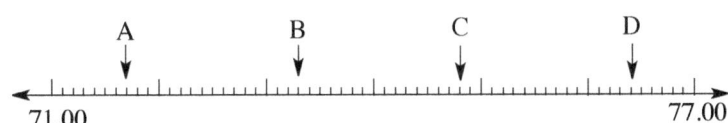

17. What is the relative position of C?

    **A.** 75.80                   **B.** 74.80
    **C.** 73.80                   **D.** 72.80

18. Which of the following is the relative position of 73.27?

    **A.** A                       **B.** B
    **C.** C                       **D.** D

19. Which of the following is 80.00 rounded to the nearest ten?

    **A.** A and B                 **B.** B and C
    **C.** C and D                 **D.** D

## 4.  Adding and Subtracting Decimals

**3–5.** Adding decimals

$$5.27 + 60.9$$

**SOLUTION**

Line up the decimal points when adding decimals.

5.27
+ 60.9

i) Line up the decimal places.
ii) Write zeros if necessary.

iii) When writing the answer, put the decimal point in the same place.

5.27
+ 60.90
.

iv) Add them as if they are whole numbers.

1
5.27
+ 60.90
66.17

added 0, but the value did not change.

**3–6.** If $x = 2.5$, what is the value of $5.7 + x$?

**SOLUTION**

| | |
|---|---|
| $5.7 + x$ | Original expression |
| $5.7 + 2.5$ | Substitute 2.5 for $x$. |
| $8.2$ | Simplify. $5.7 + 2.5 = 8.2$ |

So, the value of the expression is 8.2.

**Exercises 7**   Add the decimals.

1)   $0.31 + 0.98$                          2)   $\$2.55 + \$3$

_____              _____

3)   $22.6 + 17.4$                          4)   $7.6 + 8.5$

_____              _____

5)   $\$20.75 + \$18.85$                  6)   $2.048 + 0.77$

_____              _____

7)                              8)                              9)

$\begin{array}{r} 2 \\ + \ 4.2 \\ \hline \end{array}$     $\begin{array}{r} 10.06 \\ + \ \ 5.6 \\ \hline \end{array}$     $\begin{array}{r} \$ \ 3 \\ + \$ \ 10.10 \\ \hline \end{array}$

**Exercises 8**   Find the value of Δ.

1)   $4 + \Delta = 4.89$

2)   $24.07 + \Delta = 26.01$

_____

3)   $\Delta + 2.5 = 5.29$

4)   $\Delta + 1.81 = 3.10$

_____

**Exercises 9**   Add the decimals

1)   $2.5 + 2.8$

2)   $19.06 + 10\frac{4}{10}$

hint: $\frac{4}{10} = 0.4$

_____

3)   $\$1.05 + \$5$

4)   $4\frac{1}{2} + 2.42$

_____

5)   $\$0.33 + \$0.69$

6)   $\$34.90 + \$25.09$

_____

7)   $52.07 + 17\frac{1}{4}$

hint: $\frac{1}{4} = 0.25$

8)   $10.001 + 5\frac{4}{10}$

_____

9)   $7\frac{1}{5} + 5.55$

10)   $2.907 + 1.095$

_____

11)   $2.25 + 1\frac{3}{4}$

hint: $\frac{3}{4} = 0.75$

12)   $5.75 + 1\frac{1}{4}$

_____

**Exercises 10**    Find the value of Δ in each equation.

1)    $0.03 + Δ = 0.8$                         2)    $0.29 + Δ = 0.61$

_____                              _____

3)    $Δ + 1.05 = 3.21$                        4)    $Δ + 5.4 = 6.07$

_____                              _____

5)    $3.16 + Δ = 9.22$                        6)    $Δ + 0.42 = 1.01$

_____                              _____

7)    $Δ + 3 = 4.06$                           8)    $Δ + 2.84 = 6$

_____                              _____

9)    $1.8 + Δ = 3.55$                         10)    $Δ + 0.83 = 1.1$

_____                              _____


**Exercises 11**    Find the value using the given information.

1)    If $y = 3.2$, find the value of $y + 2$.          2)    If $y = 0.5$, find the value of $0.44 + y$.

_____                              _____

3)    If $x = 1.18$, find the value of $1.6 + x$.        4)    If $y = 0.53$, find the value of $y + 1.50$.

_____                              _____

5)    If $y - 1 = 0.67$, find the value of $y + 1.94$.    6)    If $3 + y = 6.05$, find the value of $4.26 + y$.

7)    If $2.4 + y = 3.1$, find the value of $3.1 - y$.    8)    If $y + 0.89 = 2$, find the value of $y + 1.50$.

_____                              _____

**3–7.**    Subtracting decimals

$$32.42 - 16.5$$

SOLUTION

Line up the decimal points when subtracting decimals.

$$\begin{array}{r} 32.42 \\ -\,16.5 \end{array}$$
i) Line up the decimal places
ii) Write zeros if necessary.

iii) When writing the answer, put the decimal point in the same place.

$$\begin{array}{r} 32.42 \\ -\,16.50 \end{array}$$
iv) Subtract them as if they are whole numbers.

$$\begin{array}{r} {}^{2}\overset{11}{\cancel{3}}\,{}^{14} \\ 32.42 \\ -\,16.50 \\ \hline 15.92 \end{array}$$

added 0, but the value did not change.

**Exercises 12**    Subtract the decimals.

**1)**    $63.5 - 24$                          **2)**    $11.2 - 2.9$

_____                    _____

**3)**    $\$18.05 - \$9.55$                  **4)**    $\$0.85 - \$0.25$

_____                    _____

**5)**    $50.03 - 19.87$                     **6)**    $10 - 2.85$

_____                    _____

**7)**    $\$2.45 - \$1.76$                   **8)**    $\$5.00 - \$2.18$

_____                    _____

**9)**    $\$13 - \$1.52$                     **10)**    $\$20.00 - \$8.15$

_____                    _____

**11)**
$$\begin{array}{r} 16.8 \\ -\,7 \\ \hline \end{array}$$

**12)**
$$\begin{array}{r} 22.45 \\ -\,14 \\ \hline \end{array}$$

**13)**
$$\begin{array}{r} 12.7 \\ -\,11 \\ \hline \end{array}$$

**Exercises 13**    Subtract the decimals.

    **1)**    $23.5 - 17.95$

    **2)**    $\$20.00 - \$6.75$

    **3)**    $1.75 - 0.99$

    **4)**    $\$93.90 - \$55.27$

    **5)**    $1.3 - 0.89$

    **6)**    $26.25 - 23.06$

**Exercises 14**    Find the value of $\Delta$ in each equation.

    **1)**    $2.03 - \Delta = 1.8$

    **2)**    $5.29 - \Delta = 3.61$

    **3)**    $\Delta - 2.05 = 3.21$

    **4)**    $\Delta - 1.4 = 1.07$

    **5)**    $4.16 - \Delta = 2.22$

    **6)**    $\Delta - 5.28 = 1.03$

**Exercises 15**    Find the value of each expression using the given information.

**1)**  If $y = 11.2$, find the value of $y - 9$.

**2)**  If $y = 1.05$, find the value of $3.55 - y$.

**3)**  If $x = 5.38$, find the value of $9.6 - x$.

**4)**  If $y = 4.53$, find the value of $y - 2.68$.

**5)**  If $y - 3 = 2.08$, find the value of $y - 2.94$.

**6)**  If $1 + y = 3.09$, find the value of $3.36 - y$.

**SELF-TEST**

1.  What is the value of $91.05 - 23\frac{5}{100}$?

    **A.**  70.00                    **B.**  77.66
    **C.**  67.56                    **D.**  68.00

2.  Cyndi walks 0.56 miles to school every day. After school, her mother give her a ride home, but the driving route is 1.8 times longer than what she walks. How many more miles is the ride home than the walk to school?

    **A.**  0.448                    **B.**  1.24
    **C.**  1.008                    **D.**  2.36

3.  What is the sum of $25.49 + 54.55$?

    **A.**  80.04                    **B.**  79.04
    **C.**  80.00                    **D.**  78.04

4.  At a track meet, 5 runners run the 100-meter race. Their total time is 100.8 seconds. 2 of the runners can run 2.38 meters faster than the total average. What is the sum of the time of those 2 runners?

    **A.**  45.08                    **B.**  17.78
    **C.**  35.56                    **D.**  22.52

5.  Lee began to receive an allowance of $12.50 every weekend from his parents. After four weeks he spent $34.75 to buy new shoes. How much money does he have left over?

    **A.**  $8.25                    **B.**  $10.50
    **C.**  $12.50                    **D.**  $15.25

6.  What is the sum of $2\frac{1}{5} + 2.8$?

    **A.**  5.00                    **B.**  4.62
    **C.**  4.59                    **D.**  4.52

7. It rained 2.16 inches during 3 days. On Tuesday, it rained half an inch less than the total average of rainfall. How much did it rain on Tuesday?

      **A.**  0.72                      **B.**  0.36
      **C.**  1.08                      **D.**  2.06

8. AJ has a total of $406.35 in his bank account. If he withdraws $40.00 and deposits $150.75, how much does he have in his account?

      **A.**  $81.55                 **B.**  $27.85
      **C.**  $63.05                 **D.**  $64.85

9. What is the value of Δ given the equation below?
   $$\$30.79 - \Delta = \$19.82$$

      **A.**  $10.97                 **B.**  $11.97
      **C.**  $12.97                 **D.**  $50.61

10. What is the value of Δ given the equation below?
    $$15.05 - \Delta = 9.89$$

       **A.**  5.06                    **B.**  5.16
       **C.**  6.16                    **D.**  6.06

11. What is the difference of 501.05 – 89.9?

       **A.**  411.96                **B.**  411.15
       **C.**  492.06                **D.**  421.15

12. Lily has $45.35 in her purse and pays $12.50 for two movie tickets. What is the best estimate of the remaining amount of money she has now?

       **A.**  $32.70               **B.**  $32.85
       **C.**  $33.85               **D.**  $31.85

## 5.　Multiplying and Dividing Decimals

**3–8.** Multiplying decimals

> SOLUTION
>
> a) Multiplying decimals is similar to multiplying whole numbers.
> b) Count the decimal places of the reactants before finding the product.
> c) The place of the decimal point is the sum of the decimal places of the reactants.
>
> $$\begin{array}{r} 3.6 \\ \times\ 6 \\ \end{array}$$ ←1 decimal place
>
> i) $$\begin{array}{r} \overset{3}{3.6} \\ \times\ 6 \\ \hline 216 \end{array}$$
>
> ii) $$\begin{array}{r} 3.6 \\ \times\ 6 \\ \hline 21.6 \end{array}$$ } 1 decimal place
>
> 1 decimal place
>
> i) Multiply like whole numbers.
> ii) When you find the product, move **one** decimal place to the left.
>
> So, the product of $3.6 \times 6$ is 21.6.

**3–9.** Multiplying decimals

> SOLUTION
>
> $$\begin{array}{r} 5.71 \\ \times\ 4 \\ \end{array}$$ ←2 decimal places
>
> i) $$\begin{array}{r} \overset{2}{5.71} \\ \times\ 4 \\ \hline 2284 \end{array}$$
>
> ii) $$\begin{array}{r} 5.71 \\ \times\ 4 \\ \hline 22.84 \end{array}$$ } 2 decimal places
>
> 2 decimal places
>
> i) Multiply like whole numbers.
> ii) When you find the product, move **two** decimal places to the left.
>
> So, the product of $5.71 \times 4$ is 22.84.

**Exercises 16**　Multiply the decimals.

1)　$0.2 \times 2$

2)　$0.8 \times 3$

3)　$6.5 \times 5$

4)　$2 \times 2.4$

5)　$17.6 \times 2$

6)　$15.2 \times 4$

**Exercises 17**    Multiply the decimals.

1)    $25.2 \times 2.4$

2)    $\$10.25 \times \$15$

3)    $58.4 \times 0.3$

4)    $55 \times 0.8$

5)    $7.9 \times 1.5$

6)    $5.9 \times 0.4$

**Exercises 18**    Find the value of $\Delta$ in each equation.

1)    $1.3 \times \Delta = 3.9$

2)    $4.12 \times \Delta = 12.36$

3)    $\Delta \times 0.25 = 0.55$

4)    $\Delta \times 3.4 = 10.2$

5)    $1.16 \times \Delta = 2.9$

6)    $\Delta \times 3.28 = 18.04$

**Exercises 19**    Find the value of the expression using the given information.

1)    If $y = 1.2$, find the value of $y \times 6$.

2)    If $y = 1.5$, find the value of $2.05 \times y$.

3)    If $x = 2.8$, find the value of $0.5 \times x$.

4)    If $y = 5.4$, find the value of $y \times 4.5$.

5)    If $y - 1 = 2.5$, find the value of $y \times 4.2$.

6)    If $1 + y = 3.6$, find the value of $3.4 \times y$.

**3-10.**  Multiplying decimals

SOLUTION

a) Multiply normally, ignoring the decimal points.
b) Count the number of decimal places in the reactants before finding the product.
c) The place of the decimal point in the product is the sum of the decimal places of the reactants.

$$\begin{array}{r} 2.9 \\ \times\ 6.5 \end{array} \Big\}\ \textbf{2 decimal places}$$

i) Multiply as if they are whole numbers.

ii) When you find the product, move **two** decimal places to the left.

i) $$\begin{array}{r} 2.9 \\ \times\ 6.5 \\ \hline 145 \\ +\ 174 \\ \hline 1885 \end{array}$$

ii) $$\begin{array}{r} 2.9 \\ \times\ 6.5 \\ \hline 145 \\ +\ 174 \\ \hline 18.85 \end{array} \Big\}\ \textbf{2 decimal places}$$

**2 decimal places**

So, the product of $2.9 \times 6.5$ is $18.85$.

**Exercises 20**   Multiply the decimals.

**1)**   $3.26 \times 2.0$

**2)**   $21.8 \times 3.5$

**3)**   $\$4.25 \times \$5.00$

**4)**   $\$20.40 \times \$2.42$

**5)**   $24.6 \times 6.6$

**6)**   $\$5.20 \times \$14.75$

**7)**   $31.7 \times 2.4$

**8)**   $9.55 \times 5$

**9)**   $8.28 \times 7$

**10)**   $0.75 \times 0.4$

**11)**   $2.29 \times 1.5$

**12)**   $13.1 \times 3.4$

**3–11.** Divide the decimals.

$$15 \div 0.3$$

SOLUTION

a) First, move **one** decimal place to the right for both decimals until the divisor is a whole number.

i) Move one decimal place on both.

$0.3 \overline{)15.0}$   $\longrightarrow$   $3 \overline{)150}$   * So this means that the expression of $15 \div 0.3$ equals its $150 \div 3$.

* Remember that $15. = 15.0 = 15$

b) Now solve the problem.

ii) Solve like whole numbers.

$3 \overline{)150}$   $\longrightarrow$
$$\begin{array}{r} 50 \\ 3 \overline{)150} \\ -15 \phantom{0} \\ \hline 00 \end{array}$$   $3 \times 5 = 15$

So, the quotient of $15 \div 0.3$ is 50.

**3–12.** Divide the decimals

$$1.35 \div 5$$

SOLUTION

a) First, put the decimal point of the answer in the same place as the dividend.

$5 \overline{)1.35}$   i) Line up the decimal point.   $\longrightarrow$   $5 \overline{)1.35}^{\bullet}$

b) Divide the ones digit, but 1 is not enough to divide by 5. So write down the number zero in the place as the quotient and then divide the digits in the tenth place ($13 \div 5$).

ii) 1 is not enough to divide by 5. So write down 0 in the quotient.

$5 \overline{)1.35}$   $\longrightarrow$   $5 \overline{)1.35}^{0.}$

c) Now solve the equation

iii) Solve like whole numbers.

$5 \overline{)1.35}^{0.}$   $\longrightarrow$
$$\begin{array}{r} 0.27 \\ 5 \overline{)1.35} \\ -10 \phantom{0} \\ \hline 35 \\ -35 \\ \hline 0 \end{array}$$   $5 \times 2 = 10$   $5 \times 7 = 35$

So, the quotient of $1.35 \div 5$ is 0.27.

**Exercises 21**   Divide the decimals.

1)   $9.3 \div 30.0$                          2)   $18.6 \div 2$

_____                          _____

3)   $10.05 \div 5$                          4)   $71 \div 0.5$

_____                          _____

5)   $52.5 \div 5$                           6)   $12 \div 0.2$

_____                          _____

7)   $12 \div 0.8$                           8)   $3.28 \div 8$

_____                          _____

9)   $8 \div 2.5$                            10)   $25 \div 0.5$

_____                          _____

11)   $10.8 \div 6$                          12)   $4.8 \div 0.8$

_____                          _____

13)                          14)                          15)

$4\overline{)24.64}$                $9\overline{)4.23}$                $0.3\overline{)2.37}$

**Exercises 22**   Find the value of each expression using the given information.

1)   If $y = 2.4$, find the value of $y \div 8$.        2)   If $y = 7$, find the value of $13.3 \div y$.

_____                          _____

3)   If $y = 3$, find the value of $16.8 \div y$.       4)   If $y + 2.1 = 3.7$, find the value of $y \div 4$.

_____                          _____

5)   If $y - 2.7 = 0.5$, find the value of $y \div 4$.   6)   If $1 + y = 1.9$, find the value of $0.45 \div y$.

_____                          _____

7)   If $1.01 - y = 0.20$, find the value of $y \div 9$.  8)   If $2(y + 0.20) = 0.84$, find the value of $y \div 0.5$.

_____                          _____

**3-13.**  Divide the decimals.

$$2.16 \div 0.4$$

**SOLUTION**

a) First, move **one** decimal place to the right for both decimals until the divisor is a whole number.

i) move one decimal place on both.

$0.4\overline{)2.16}$  $\longrightarrow$  $4\overline{)21.6}$

\* Remember that 4. = 4.0 = 4

\* So this means that the expression of $2.16 \div 0.4$ equals its $21.6 \div 4$.

b) Second, write the decimal point in the same place as the dividend.

$4\overline{)21.6}$  ii) line up the decimal point.  $\longrightarrow$  $4\overline{)21.6}$

c) Now solve the problem.

$4\overline{)21.6}$  iii) Solve like whole numbers.  $\longrightarrow$

$$\begin{array}{r} 5.4 \\ 4\overline{)21.6} \\ -20 \quad\quad 4 \times 5 = 20 \\ \hline 16 \\ -16 \quad\quad 4 \times 4 = 16 \\ \hline 0 \end{array}$$

So, the quotient of $2.16 \div 0.4$ is 5.4.

**Exercises 23**    Divide the decimals.

**1)**    $2.193 \div 0.5$

**2)**    $13.44 \div 2.1$

**3)**    $26.468 \div 5.2$

**4)**    $0.119 \div 0.5$

**5)**    $1.656 \div 4.6$

**6)**    $1.278 \div 0.6$

**Exercises 24**    Divide the decimals.

1)    $634.6 \div 9.5$

2)    $32.634 \div 6.3$

3)    $29.7 \div 4.5$

4)    $12.6 \div 0.6$

5)    $0.07\overline{)0.1561}$

6)    $7.14\overline{)9.5676}$

7)    $1.02\overline{)5.8242}$

**Exercises 25**    Find the value of Δ in each equation.

1)    $2.814 \div \Delta = 2.01$

2)    $0.312 \div \Delta = 0.65$

3)    $\Delta \div 3.5 = 4.41$

4)    $\Delta \div 1.61 = 5.6$

5)    $4.27 \div \Delta = 3.05$

6)    $\Delta \div 2.4 = 5.65$

**Exercises 26**    Find the value of each expression using the given information.

1)    If $y = 4.32$, find the value of $y \div 0.5$.

2)    If $y = 2.9$, find the value of $4.35 \div y$.

3)    If $x = 3.6$, find the value of $28.8 \div x$.

4)    If $y = 5.43$, find the value of $y \div 1.5$.

5)    If $y - 1.9 = 1.04$, find the value of $y \div 3.5$.

6)    If $7.51 + y = 13.6$, find the value of $10.962 \div y$.

SELF-TEST

1. Find the sum of the expression below.
   4.09 + 4.405

   A. 8.495            B. 8.414
   C. 9.305            D. 8.404

2. Which of the following is the closest estimated product of 6.45 × 9.4?

   A. 60.63            B. 54.00
   C. 15.85            D. 58.30

3. Find the product of the expression below.
   27.2 × 2.05

   A. 5.576            B. 53.76
   C. 54.76            D. 55.76

4. Find the quotient of the expression below.
   34.02 ÷ 5.4

   A. 6.18             B. 6.3
   C. 63               D. 0.63

5. Ivan has $7.20 to buy candy. If 3 chocolate bars costs $0.48, how many chocolate bars can he buy?

   A. 15               B. 30
   C. 45               D. 60

6. Find the difference of the expression below.
   64.2 − 5.92

   A. 0.5              B. 58.10
   C. 58.28            D. 58.12

7. It rained 0.14 inches on Monday. On Tuesday, it rained 0.11 inches less than on Monday. What is the total sum of rainfall on both days?

    **A.** 0.0.28                                      **B.** 0.03
    **C.** 0.25                                        **D.** 0.17

8. Justin has $26.85 on his hand. He bought a movie ticket that cost $13.95. Additionally, he bought two small bags of popcorn that cost $4.45 each. How much money will he have left over?

    **A.** $4.00                                       **B.** $8.00
    **C.** $18.40                                      **D.** $22.85

9. What is the value of Δ for the equation below?
$$\$6.50 \times \Delta = \$29.25$$

    **A.** $4.05                                       **B.** $4.50
    **C.** $5.50                                       **D.** $3.05

10. What is the value of Δ for the equation below?
$$41.4 \div \Delta = 6$$

    **A.** 74.40                                       **B.** 35.40
    **C.** 246.60                                      **D.** 6.90

11. What is the value of Δ for the equation below?
$$\Delta \div \$2.05 = \$8.00$$

    **A.** $16.85                                      **B.** $16.04
    **C.** $16.00                                      **D.** $16.40

12. What is the value of Δ for the equation below?
$$\Delta \times 2.50 = 14.00$$

    **A.** 35.00                                       **B.** 11.50
    **C.** 5.60                                        **D.** 16.50

## 6.  Understanding Fractions

**3–14.**  Write a fraction for the shaded part of the circle.

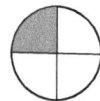

**SOLUTION**

The circle is divided into four equal parts and contains one shaded part.

$$\frac{\text{Number of parts in the shaded region}}{\text{Number of equal parts in all}} = \frac{1 \leftarrow \text{numerator}}{4 \leftarrow \text{denominator}}$$

Read as = "one divided by four", "one out of four", or "one fourth".

---

**Exercises 27**    Write each fraction in its word form.

1) $\dfrac{13}{100} =$

_____

2) $2\dfrac{7}{8} =$

_____

3) $2\dfrac{4}{5} =$

_____

4) $\dfrac{89}{100} =$

_____

**Exercises 28**    Write each expression in its standard form.

1)  Four ninths _____

2)  Ten and two divided by seven _____

3)  Six out of eleven _____

4)  Three and five divided by twelve subtracted by six divided by eight

_____

**3–15.** Find the LCD of $\frac{7}{12}$ and $\frac{1}{8}$.

SOLUTION

The least common denominator (LCD) is the smallest positive integer that can be divided by the given denominators. The LCD is the least common multiple (LCM) of the denominators of the fractions.

a) First, find the LCM of the denominators.

$\frac{7}{12}$ and $\frac{1}{8}$

**Denominators**

⎰ Multiples of 12: 12, **24**, 36, and so on.
⎱ Multiples of 8: 8, 16, **24**, 32, and so on.

The LCM of the denominators, 12 and 8, is **24**.

b) The LCM of the denominators is also the LCD.

So, the LCD of $\frac{7}{12}$ and $\frac{1}{8}$ is 24.

## 7.  Comparing Fractions

**3–16.** Compare the fractions.

$$\frac{2}{9} \; \square \; \frac{5}{9}$$

SOLUTION

Given that two different fractions have the <u>same denominators</u>, then compare the numerators in order to compare the fractions.

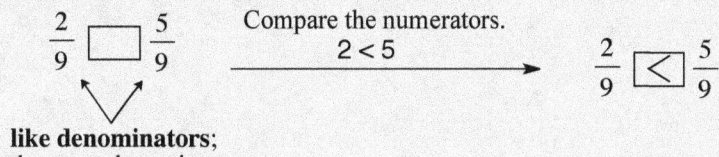

$\frac{2}{9} \; \square \; \frac{5}{9}$     Compare the numerators.    $2 < 5$ ⟶    $\frac{2}{9} \; \boxed{<} \; \frac{5}{9}$

**like denominators;**
the same denominators

**3–17.**   If two different fractions have <u>different denominators</u>, then follow the direction in order to compare the fractions.

a.  First, find the LCD of both fractions.
b.  Second, rewrite the fractions as equivalent.
c.  Then, compare the numerators of the fractions.

**3–18.** Compare the fractions.

$$\frac{1}{2} \ \square \ \frac{1}{4}$$

**SOLUTION**

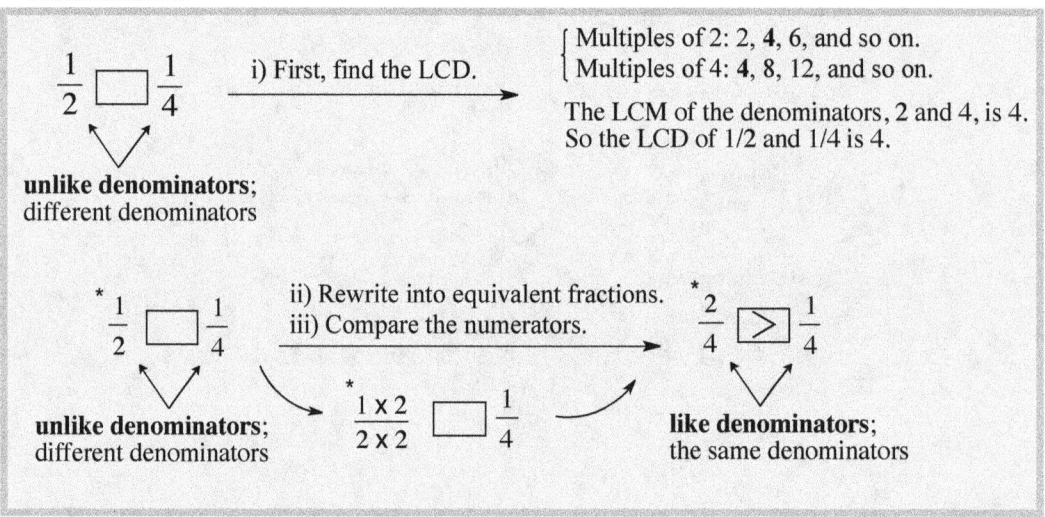

\* Equivalent fractions: The fractions are equivalent.

$$\frac{1}{2} \ = \ \frac{1 \times 2}{2 \times 2} \ = \ \frac{2}{4}$$

equivalent fractions

**Exercises 29**   Compare the fractions. Use the symbols of < (less than), = (equal to), > (greater than).

1)  $\frac{1}{7} \ \square \ \frac{3}{7}$    2)  $\frac{1}{3} \ \square \ \frac{2}{3}$

3)  $\frac{5}{7} \ \square \ \frac{4}{7}$    4)  $\frac{2}{5} \ \square \ \frac{3}{5}$

5)  $\frac{1}{4} \ \square \ \frac{1}{5}$    6)  $\frac{7}{8} \ \square \ \frac{7}{9}$

7)  $\frac{1}{2} \ \square \ \frac{1}{3}$    8)  $\frac{10}{11} \ \square \ 1$

**Exercises 30**   Compare the fractions. Use the symbols of < (less than), = (equal to), > (greater than).

1)  $\dfrac{2}{8}$ $\square$ $\dfrac{2}{4}$

2)  $\dfrac{4}{8}$ $\square$ $\dfrac{1}{4}$

3)  $\dfrac{2}{4}$ $\square$ $\dfrac{3}{6}$

4)  $\dfrac{5}{10}$ $\square$ $\dfrac{10}{20}$

5)  $\dfrac{5}{6}$ $\square$ $\dfrac{5}{7}$

6)  $\dfrac{2}{3}$ $\square$ $\dfrac{3}{4}$

7)  $\dfrac{5}{2}$ $\square$ $1\dfrac{1}{2}$

8)  $\dfrac{3}{2}$ $\square$ $1$

9)  $\dfrac{2}{3}$ $\square$ $3$

10)  $\dfrac{3}{4}$ $\square$ $\dfrac{1}{2}$

**Exercises 31**   List from greatest to least.

1)  $2\dfrac{1}{4}$, $2\dfrac{1}{6}$, $2\dfrac{1}{2}$, $2\dfrac{1}{5}$, $2\dfrac{1}{3}$

2)  $2.85$, $2\dfrac{2}{3}$, $2\dfrac{3}{4}$, $2\dfrac{4}{5}$, $2\dfrac{5}{6}$

_____

3)  $2\dfrac{1}{10}$, $2\dfrac{11}{100}$, $2\dfrac{3}{5}$, $2\dfrac{33}{50}$, $2\dfrac{16}{25}$

4)  $2\dfrac{1}{4}$, $2\dfrac{4}{20}$, $2\dfrac{45}{100}$, $2\dfrac{15}{40}$, $2.24$

_____

5)  $\dfrac{1}{4}$, $\dfrac{10}{15}$, $\dfrac{7}{10}$, $\dfrac{1}{5}$, $\dfrac{9}{30}$

6)  $\dfrac{2}{7}$, $0.34$, $\dfrac{1}{4}$, $0.306$, $\dfrac{3}{9}$

_____

**Exercises 32**    Use the number line to find the unknown values.

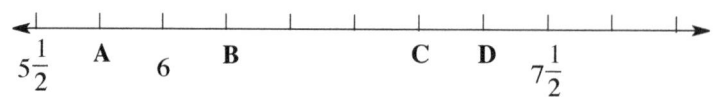

**1)**  A = _____

**2)**  B = _____

**3)**  C = _____

**4)**  D = _____

**Exercises 33**    Use the number line to determine the unknown values.

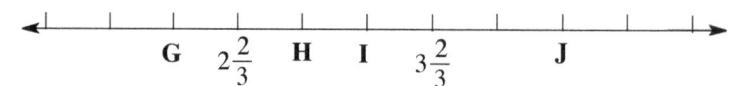

**1)**  G = _____

**2)**  H = _____

**3)**  I = _____

**4)**  J = _____

**Exercises 34**    Use the number line to find the unknown values.

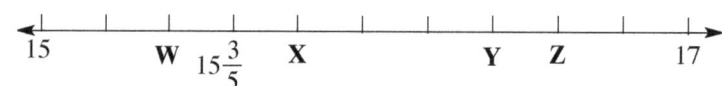

**1)**  W = _____

**2)**  X = _____

**3)**  Y = _____

**4)**  Z = _____

**Exercises 35**    Mark each letter on the number line.

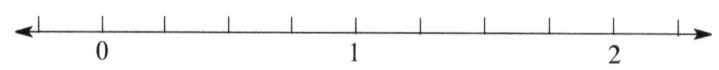

**1)**  S = $\frac{1}{3}$

**2)**  T = $\frac{7}{10}$

**3)**  U = $1\frac{2}{5}$

**4)**  V = $1\frac{5}{6}$

SELF-TEST

1.  Which of the following is the best description in word form?

$$\frac{7}{19}$$

    **A.** Seven and nineteen tenths     **B.** Seven and nineteen hundredths
    **C.** Seven divided by nineteen hundredths **D.** Seven out of nineteen

2.  Which of the following best describes the decimal below?
        Thirty-seven and two hundred three thousandths

    **A.** $37\frac{203}{100}$                **B.** $372\frac{3}{1000}$
    **C.** 37.203               **D.** 372.03

3.  Which of the following is equivalent to the decimal below?
        One and five thousandths

    **A.** $1\frac{5}{1000}$                **B.** $\frac{1}{500}$
    **C.** 1.500                **D.** 0.105

4.  Which of the following is the prime factorization for 8?

    **A.** $1 \times 2 \times 2 \times 4$         **B.** $2 \times 4$
    **C.** $2 \times 2 \times 2$           **D.** $1 \times 8$

5.  What is the common factor of 27 and 39?

    **A.** 3                **B.** 2 and 3
    **C.** 2                **D.** 1

6.  What is the LCM of 5 and 7?

    **A.** 20               **B.** 30
    **C.** 35               **D.** 40

7.  What is the greatest common factor of 24 and 38?

    **A.** 2                **B.** 3
    **C.** 5                **D.** 7

**8.** Which of the following is the prime factorization of 84?

    **A.** $2 \times 2 \times 3 \times 7$               **B.** $12 \times 7$

    **C.** $4 \times 3 \times 7$                 **D.** $4 \times 21$

**9.** What is the LCD of $\dfrac{8}{12}$ and $\dfrac{1}{8}$?

    **A.** 8                       **B.** 12

    **C.** 36                    **D.** 24

\* Use the number line for Exercises **10-12**.

**10.** What is the relative position of **A**?

    **A.** $\dfrac{1}{5}$               **B.** $\dfrac{1}{7}$

    **C.** $\dfrac{1}{6}$               **D.** $\dfrac{1}{8}$

**11.** Which of the following is the relative position of $1\dfrac{3}{7}$?

    **A.** A                **B.** B

    **C.** C                **D.** D

**12.** What is the relative position of **B**?

    **A.** $\dfrac{2}{5}$               **B.** $\dfrac{5}{6}$

    **C.** $\dfrac{4}{7}$               **D.** $\dfrac{5}{8}$

\* Use the number line for Exercises **13-16**.

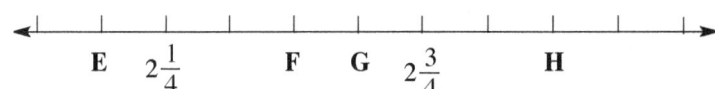

13. What is the relative position of **E**?

      **A.** $1\dfrac{1}{2}$                      **B.** $1\dfrac{1}{8}$

      **C.** $2\dfrac{1}{2}$                      **D.** $2\dfrac{1}{8}$

14. Which of the following is the relative position of $2\dfrac{1}{2}$?

      **A.** E                      **B.** F
      **C.** G                      **D.** H

15. What is the relative position of **H**?

      **A.** $2\dfrac{7}{8}$                      **B.** $2\dfrac{3}{4}$

      **C.** 3                       **D.** $3\dfrac{1}{8}$

16. Which of the following is 2 rounded to the nearest tenth?

      **A.** E                      **B.** E and F
      **C.** E, F, and G             **D.** H

17. Which of the following correctly lists the numbers from least to greatest?
$$1\dfrac{6}{10},\ 1\dfrac{3}{6},\ 1\dfrac{2}{5},\ 1.67,\ 1.55$$

      **A.** $1\dfrac{2}{5} < 1.55 < 1\dfrac{3}{6} < 1\dfrac{6}{10} < 1.67$     **B.** $1\dfrac{3}{6} < 1\dfrac{2}{5} < 1.55 < 1\dfrac{6}{10} < 1.67$

      **C.** $1\dfrac{2}{5} < 1\dfrac{3}{6} < 1.55 < 1.67 < 1\dfrac{6}{10}$     **D.** $1\dfrac{2}{5} < 1\dfrac{3}{6} < 1.55 < 1\dfrac{6}{10} < 1.67$

18. Which of the following correctly lists the numbers from least to greatest?
$$\dfrac{1}{3},\ \dfrac{2}{5},\ \dfrac{3}{7},\ \dfrac{4}{9},\ \dfrac{5}{12}$$

      **A.** $\dfrac{1}{3} > \dfrac{2}{5} > \dfrac{3}{7} > \dfrac{4}{9} > \dfrac{5}{12}$       **B.** $\dfrac{1}{3} < \dfrac{2}{5} < \dfrac{3}{7} < \dfrac{4}{9} < \dfrac{5}{12}$

      **C.** $\dfrac{1}{3} < \dfrac{1}{4} < \dfrac{3}{8} < \dfrac{5}{12} < \dfrac{3}{6}$       **D.** $\dfrac{1}{3} < \dfrac{2}{5} < \dfrac{5}{12} < \dfrac{3}{7} < \dfrac{4}{9}$

## 8.  Change Decimals and Fractions
**3–19.**  Change with fractions

> Convert the decimals to their fraction form and the fractions to their decimal forms.
>
> $$0.1 = \frac{1}{10}, \quad 0.01 = \frac{1}{100}, \quad 0.001 = \frac{1}{1000}, \quad 1.1 = \frac{11}{10} = 1\frac{1}{10}$$

**Exercises 36**    Write each decimal as a fraction.

**1)**  2.5                                    **2)**  0.75

_____            _____

**3)**  0.39                                   **4)**  0.10

_____            _____

**5)**  1.2                                    **6)**  1.05

_____            _____

**7)**  2.05                                   **8)**  0.004

_____            _____

**9)**  0.009                                  **10)**  20.101

_____            _____

**Exercises 37**    Write each fraction as a decimal.

**1)**  $\frac{1}{8}$                          **2)**  $\frac{3}{12}$

_____            _____

**3)**  $\frac{2}{5}$                          **4)**  $\frac{4}{5}$

_____            _____

**5)**  $\frac{6}{15}$                         **6)**  $2\frac{1}{5}$

_____            _____

**7)**  $\frac{7}{16}$                         **8)**  $5\frac{9}{12}$

_____            _____

## 9.  Adding and Subtracting Fractions

**3–20.**  Adding Fractions

$$\frac{1}{9} + \frac{4}{9}$$

**SOLUTION**

You can add fractions with **like denominators** by adding the numerators.

Add numerators like whole numbers.

$$\frac{1}{9} + \frac{4}{9} = \frac{1+4}{9} = \frac{5}{9}$$

Keep the same denominators.

**like denominators**;
the same denominators

So, the sum of $\frac{1}{9} + \frac{4}{9}$ is $\frac{5}{9}$.

**3–21.**  Add the values.

$$0.5 + \frac{3}{4}$$

**SOLUTION**

You can convert the decimal as a fraction and then compare the denominators. If fractions are unlike denominators in the expression, then find the **Least Common Denominators (LCD)**. If fractions have like denominators, then just add the numerators together to find the sum.

$$0.5^* + \frac{3}{4} = \frac{2}{4} + \frac{3}{4} = \frac{2+3}{4} = \frac{5}{4} \text{ or } 1\frac{1}{4}$$

**like denominators**;
the same denominators

$$0.5^* = \frac{5}{10} = \left[\frac{5 \times 1}{5 \times 2}\right] = \frac{1}{2} = \frac{2}{4}$$

equivalent values

So, the sum of $0.5 + \frac{3}{4}$ is $1\frac{1}{4}$.

**Exercises 38**    Find the sum of each expression.

1)    $\dfrac{6}{5} + \dfrac{1}{5}$

2)    $2\dfrac{7}{8} + 1\dfrac{3}{8}$

3)    $\dfrac{3}{5} + 3.4$

4)    $4\dfrac{6}{7} + 1\dfrac{2}{7}$

5)    $2\dfrac{1}{5} + \dfrac{2}{5}$

6)    $4\dfrac{1}{6} + \dfrac{5}{6}$

7)    $\dfrac{3}{10} + \dfrac{4}{10}$

8)    $0.25 + \dfrac{6}{8}$

9)    $1\dfrac{2}{3} + 2\dfrac{5}{3}$

10)    $0.4 + \dfrac{4}{10}$

11)    $\dfrac{1}{4} + \left(\dfrac{1}{2} + 1\dfrac{1}{2}\right)$

12)    $\left(\dfrac{2}{3} + 1\dfrac{1}{3}\right) + \dfrac{1}{8}$

**Exercises 39**    Find the value using the given information.

1)    If $y = 3\dfrac{5}{6}$, find the value of $y + 2$.

2)    If $y = 1\dfrac{5}{6}$, find the value of $\dfrac{5}{6} + y$.

3)    If $y - \dfrac{2}{3} = 2$, find the value of $3 + (y - 2)$.

4)    If $y - 8 = 10\dfrac{1}{2}$, find the value of $y - 10\dfrac{1}{2}$.

5)    If $y - 1 = \dfrac{3}{4}$, find the value of $y - \dfrac{3}{4}$.

6)    If $y - 2 = \dfrac{2}{5}$, find the value of $y - \dfrac{2}{5}$.

**3-22.**  Adding fractions

$$\frac{5}{6} + \frac{1}{2}$$

**SOLUTION**

If fractions are **unlike denominators** in the expression, then find the Least Common Denominators (LCD).

$$\frac{5}{6} + \frac{1}{2}$$   i) First, find the LCD.   $\begin{cases} \text{Multiples of 6: } \mathbf{6}, 12, 18, \text{ and so on.} \\ \text{Multiples of 2: } 2, 4, \mathbf{6}, 8, \text{ and so on.} \end{cases}$

The LCM of 6 and 2 is **6**.

**unlike denominators**;   The LCD of $\frac{5}{6}$ and $\frac{1}{2}$ is 6.
different denominators

ii) Second, rewrite the fractions as equivalent fractions with a LCD.
iii) Now you can add fractions with **like denominators** and then simplify it.

$$\frac{5}{6} + \frac{1}{2} = \frac{5}{6} + \frac{1 \times 3}{2 \times 3} = \frac{5}{6} + \frac{3}{6} = \frac{8}{6} \text{ or } 1\frac{1}{3}$$

**Unlike denominators** $\xrightarrow{\text{ii)}}$ **like denominators** $\xrightarrow{\text{iii)}}$ simplify.

So, the sum of $\frac{5}{6} + \frac{1}{2}$ is $1\frac{1}{3}$.

**3-23.**  Add the fractions.

$$\frac{5}{6} + 1\frac{1}{2}$$

**SOLUTION**

When the fractions have mixed numbers with a whole number, first convert it to an improper fraction. If the fractions have **unlike denominators**, then find the Least Common Denominators (LCD).

$$\frac{5}{6} + 1\frac{1}{2} = \frac{5}{6} + \frac{3}{2}$$

i) Change the mixed numbers
to improper fractions.*

ii) Rewrite the fractions as equivalent with a LCD. The LCD of $\frac{5}{6}$ and $1\frac{1}{2}$ is **6**.

iii) Now you can add fractions with **like denominators** and simplify if necessary.

$$\frac{5}{6} + \frac{3}{2} = \frac{5}{6} + \frac{3 \times 3}{2 \times 3} = \frac{5}{6} + \frac{9}{6} = \frac{5+9}{6} = \frac{14}{6} \text{ or } 2\frac{1}{3}$$

Unlike denominators $\xrightarrow{\text{ii)}}$ like denominators $\xrightarrow{\text{iii)}}$ simplify.

So, the sum of $\frac{5}{6} + 1\frac{1}{2}$ is $2\frac{1}{3}$.

* Improper fraction: A fraction with a numerator that is greater than or equal to the denominator in the fraction. For example; $\frac{3}{1}$, $\frac{5}{2}$, or $\frac{3}{3}$

**Exercises 40**   Add the fractions and simplify if necessary.

1)  $\frac{6}{7} + \frac{1}{5}$

2)  $2\frac{1}{3} + \frac{2}{9}$

3)  $2\frac{1}{2} + 1\frac{3}{4}$

4)  $4\frac{2}{5} + 1\frac{2}{4}$

5)  $1\frac{3}{6} + \frac{3}{4}$

6)  $2\frac{2}{4} + 1\frac{3}{6}$

7)  $4\frac{1}{2} + \frac{5}{8}$

8)  $2\frac{2}{3} + \frac{3}{5}$

9)  $1\frac{1}{2} + 2\frac{2}{3}$

10)  $\frac{1}{4} + 2\frac{5}{6}$

11)  $\frac{1}{3} + 3\frac{3}{5}$

12)  $\frac{4}{7} + 1\frac{3}{4}$

**Exercises 41**    Add the fractions and simplify if necessary.

1)    $1\frac{1}{7} + 2\frac{2}{3}$

2)    $2\frac{2}{3} + 1\frac{3}{8}$

3)    $4\frac{1}{2} + \frac{1}{4}$

4)    $2\frac{2}{4} + \frac{3}{5}$

5)    $1\frac{1}{4} + \frac{1}{7}$

6)    $2\frac{7}{9} + \frac{5}{6}$

7)    $\left(\frac{1}{2} + 1\frac{1}{3}\right) + 2\frac{1}{6}$

8)    $\left(\frac{3}{4} + 1\frac{1}{5}\right) + \frac{1}{20}$

**Exercises 42**    Find the value of Δ.

1)    $\Delta + \frac{2}{5} = \frac{3}{5}$

2)    $\Delta + 1 = 1\frac{5}{7}$

3)    $\Delta + \frac{5}{6} = 1\frac{1}{3}$

4)    $\Delta + \frac{1}{2} = 2$

5)    $1\frac{1}{2} + \Delta = 3$

6)    $\frac{1}{3} + \Delta = \frac{1}{2}$

7)    $\frac{3}{4} + \Delta = 1\frac{1}{2}$

8)    $\Delta + \frac{5}{6} = 1\frac{5}{12}$

**3–24.**   Subtracting Fractions

$$\frac{5}{7} - \frac{2}{7}$$

**SOLUTION**

Subtracting fractions is similar to adding fractions.

Subtract numerators like whole numbers.

$$\frac{5}{7} - \frac{2}{7} = \frac{5-2}{7} = \frac{3}{7}$$

Keep the same denominators.

**like denominators**;
the same denominators

So, the difference of $\frac{5}{7} - \frac{2}{7}$ is $\frac{3}{7}$.

**Exercises 43**    Find the value of each expression and simplify if necessary.

1)  $\dfrac{6}{7} - \dfrac{1}{7}$

2)  $\dfrac{9}{15} - \dfrac{2}{15}$

3)  $4\dfrac{1}{8} - 4$

4)  $2\dfrac{1}{5} - \dfrac{2}{5}$

5)  $2 - \dfrac{4}{10}$

6)  $2\dfrac{7}{8} - 1\dfrac{3}{8}$

7)  $4\dfrac{6}{7} - 1\dfrac{2}{7}$

8)  $1 - \dfrac{2}{7}$

9)  $2\dfrac{1}{6} - \dfrac{5}{6}$

10)  $4\dfrac{2}{3} - 2\dfrac{5}{3}$

11)  $1\dfrac{1}{3} - \left(1\dfrac{1}{2} - \dfrac{1}{2}\right)$

12)  $\dfrac{2}{3} - \left(2 - 1\dfrac{2}{6}\right)$

**3-25.** Subtracting fractions

$$1\frac{1}{8} \ - \ \frac{1}{2}$$

**SOLUTION**

If they are unlike fractions, then 1) find the LCD, 2) rewrite the fractions as equivalent fractions with a LCD, and 3) subtract the fractions.

$$1\frac{1}{8} \ - \ \frac{1}{2} \ = \ \frac{9}{8} \ - \ \frac{1}{2}$$

i) Change the mixed numbers to improper fractions.

ii) Rewrite the fractions as equivalent with a LCD. The LCD of $1\frac{1}{8}$ and $\frac{1}{2}$ is **8**.

iii) Now you can subtract fractions with **like denominators** and simplify if necessary.

$$\frac{9}{8} \ - \ \frac{1}{2} \ = \ \frac{9}{8} \ - \ \frac{1 \times 4}{2 \times 4} \ = \ \frac{9}{8} \ - \ \frac{4}{8} \ = \ \frac{9-4}{8} \ = \ \frac{5}{8}$$

   **unlike denominators**        **like denominators**

So, the difference of $1\frac{1}{8} - \frac{1}{2}$ is $\frac{5}{8}$.

**Exercises 44**    Subtract the fractions and simplify if necessary.

1)  $3\frac{1}{2} \ - \ 2\frac{1}{6}$

2)  $2\frac{2}{9} \ - \ \frac{1}{3}$

3)  $5\frac{7}{4} \ - \ \frac{5}{8}$

4)  $3\frac{1}{2} \ - \ 1\frac{1}{7}$

5)  $1\frac{1}{3} \ - \ 1\frac{1}{6}$

6)  $3\frac{1}{8} \ - \ 2\frac{1}{2}$

7)  $3\frac{1}{3} \ - \ \frac{3}{5}$

8)  $2\frac{1}{4} \ - \ \frac{5}{6}$

**Exercises 45**    Subtract the fractions and simplify if necessary.

1)    $3\frac{2}{3} - 1\frac{1}{7}$          2)    $2\frac{1}{4} - 1\frac{2}{3}$

3)    $2\frac{1}{5} - \frac{1}{4}$          4)    $1\frac{1}{2} - \frac{4}{9}$

5)    $1\frac{1}{2} - \frac{2}{3}$          6)    $1\frac{1}{3} - 1\frac{1}{5}$

7)    $1\frac{4}{5} - \frac{3}{8}$          8)    $2\frac{2}{5} - 1\frac{1}{6}$

9)    $1\frac{7}{9} - \frac{5}{6}$          10)    $3\frac{1}{7} - \frac{3}{8}$

**Exercises 46**    Find the value of $\Delta$ in each equation.

1)    $\Delta - \frac{1}{2} = \frac{5}{6}$          2)    $\Delta - 1 = 1\frac{3}{4}$

3)    $\Delta - \frac{2}{7} = 1\frac{1}{5}$          4)    $\Delta - \frac{1}{3} = 3$

5)    $1\frac{1}{2} - \Delta = 1$          6)    $\frac{3}{5} - \Delta = \frac{1}{2}$

7)    $1\frac{3}{8} - \Delta = \frac{1}{2}$          8)    $\Delta - \frac{5}{12} = \frac{1}{3}$

# * Solving Problems

**Exercises 47**   Solve each problem using the given information.

Use the information below for Exercises **1-6**. Mrs. Suzan divided 60 people in 5 equal groups for the school activities. $\frac{5}{12}$ of the total number of people are girls and $\frac{1}{12}$ are boys. The rest are adults. Show your answer.

1) Name the fraction that represents the sum of girls and adults.

2) Name the fraction that represents the difference of boys and girls.

3) How many boys are in the group? Show your answer.

4) How many people are in each group? Show your answer.

5) How many girls are in each group if Mrs. Suzan decides to put an equal number of girls in each group? Show your answer.

6) How many adults are in each group if there is an equal number in each group? Show your answer.

SELF-TEST

1. Find the value of the following expression.

$$\frac{3}{5} - \frac{4}{10}$$

A. $\frac{2}{5}$                    B. $\frac{1}{10}$

C. $\frac{1}{5}$                    D. $\frac{3}{10}$

\* Use the information below for Exercises **2-7**. There are 225 Lego pieces in a storage box. The Legos come in red, yellow, brown, and green. The number of brown pieces is one and half the number of red pieces and there are 15 more green pieces than yellow pieces. $\frac{8}{45}$ of the Lego pieces are green.

**2.** How many red Legos are there?

    **A.** 35               **B.** 40

    **C.** 45               **D.** 53

**3.** What fraction is the sum of red and yellow pieces?

    **A.** 85               **B.** 95

    **C.** 98               **D.** 102

**4.** How many yellow pieces are there?

    **A.** 30               **B.** 40

    **C.** 45               **D.** 50

**5.** What fraction is the difference between red and brown pieces?

    **A.** 18               **B.** 20

    **C.** 22               **D.** 24

**6.** If you divide the Legos in piles of 15, how many piles are there?

    **A.** 15               **B.** 20

    **C.** 25               **D.** 30

**7.** How many green pieces are there?

    **A.** 18               **B.** 20

    **C.** 35               **D.** 24

**8.** Which of the following is the sum of $\frac{2}{7}+\frac{3}{4}$?

    **A.** $\frac{5}{7}$               **B.** $\frac{5}{4}$

    **C.** $1\frac{1}{28}$           **D.** $\frac{5}{12}$

* Use the information below for Exercises **9-10**. At the aquarium, the orcas are fed 50 pounds of fish. The orcas are fed $1\frac{1}{2}$ as many pounds than the dolphins.

9.  How many pounds of fish do the dolphins eat?

   A.  33                              B.  25
   C.  50                              D.  75

10.  If the seals are fed $\frac{7}{8}$ as much as what the orcas are fed, what fraction is the difference between the feeding amount of the seals and dolphins?

   A.  18                              B.  10.45
   C.  43.75                           D.  52.25

11.  What is the sum of $2\frac{2}{3} + 1\frac{2}{6}$ ?

   A.  $3\frac{4}{6}$                  B.  $4\frac{1}{3}$
   C.  3                               D.  4

12.  What is the value of $\Delta$ for the equation below?
$$1\frac{1}{3} - \Delta = 1$$

   A.  $\frac{2}{3}$                   B.  $\frac{1}{6}$

   C.  $1\frac{1}{3}$                  D.  $\frac{1}{3}$

13.  What is the value of $\Delta$ for the equation below?
$$2 - \Delta = \frac{2}{3}$$

   A.  $1\frac{1}{6}$                  B.  $1\frac{1}{3}$

   C.  $\frac{1}{6}$                   D.  $\frac{2}{3}$

## 10.  Multiplying and Dividing Fractions

**3–26.**  Multiplying with fractions.

$$\frac{1}{2} \times \frac{1}{3}$$

**SOLUTION**

Multiplying fractions is similar to multiplying whole numbers.

Multiply the numerators together.

$$\frac{1}{2} \times \frac{1}{3} = \frac{1}{2} \times \frac{1}{3} = \frac{1 \times 1}{2 \times 3} = \frac{1}{6}$$

Multiply the dedominators together.

So, the product of $\frac{1}{2}$ and $\frac{1}{3}$ is $\frac{1}{6}$.

**3–27.**  Multiply the fractions.

$$\frac{6}{7} \times \frac{1}{4}$$

**SOLUTION**

If the numerators and the denominators in the expression have a common factor, you can simplify before multiplying.

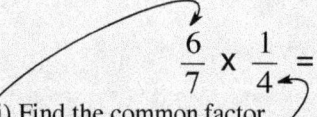

$$\frac{6}{7} \times \frac{1}{4} =$$

i) Find the common factor of 6 and 4.
ii) Divide by the common factor.

$\begin{cases} \text{The factor of 6: } \mathbf{2} \times 3 = 6 \\ \text{The factor of 4: } \mathbf{2} \times 2 = 4 \end{cases}$

The common factor of 6 and 4 is **2**.

So, 6 and 4 can be divided by 2.

Multiply the numerators together.

$$\frac{6}{7} \times \frac{1}{4} = \frac{\overset{3}{6}}{7} \times \frac{1}{\underset{2}{4}} = \frac{3 \times 1}{7 \times 2} = \frac{3}{14}$$

The common factor of 6 and 4 is **2**.

Multiply the denominators together.

**Exercises 48**  Multiply the fractions.

1) $2\frac{1}{6} \times \frac{2}{3}$

2) $1\frac{2}{3} \times 2\frac{1}{2}$

_____

_____

3) $\frac{4}{7} \times \frac{3}{4}$

4) $\frac{1}{3} \times \frac{3}{5}$

_____

_____

5) $1\frac{1}{4} \times \frac{2}{5}$

6) $\frac{2}{3} \times \frac{5}{8}$

_____

_____

7) $2 \times \frac{4}{10}$

8) $2\frac{2}{3} \times 1\frac{1}{5}$

_____

_____

9) $3\frac{1}{7} \times \frac{2}{8}$

10) $4\frac{1}{8} \times 4$

_____

_____

**Exercises 49**   Find the value using the given information.

1) If $y = \frac{5}{6}$, find the value of $y \times 2$.

2) If $y = 2\frac{2}{5}$, find the value of $\frac{5}{6} \times y$.

_____

_____

3) If $y \times \frac{2}{3} = 2$, find the value of $3 \times y - 2$.

4) If $\frac{3}{4} - y = \frac{1}{2}$, find the value of $y \times 1\frac{1}{3}$.

_____

_____

5) If $y - 1 = \frac{3}{4}$, find the value of $y \times \frac{4}{7}$.

6) If $2 + y = 3\frac{1}{5}$, find the value of $y \times \frac{5}{16}$.

_____

_____

**Exercises 50**  Multiply the fractions and simplify if necessary.

1) $3\frac{1}{2} \times 1\frac{1}{7}$

2) $3\frac{1}{8} \times 2\frac{1}{2}$

3) $1\frac{1}{3} \times 1\frac{1}{6}$

4) $\frac{2}{9} \times \frac{1}{3}$

5) $2\frac{1}{2} \times 1\frac{3}{5}$

6) $\frac{4}{5} \times \frac{3}{8}$

7) $1\frac{1}{2} \times \frac{2}{3}$

8) $3\frac{2}{3} \times 1\frac{2}{7}$

9) $2\frac{1}{4} \times 1\frac{2}{3}$

10) $2\frac{2}{7} \times 1\frac{3}{4}$

**Exercises 51**  Find the value of Δ.

1) $\Delta \times \frac{1}{2} = 2$

2) $\Delta \times 2 = \frac{3}{4}$

3) $\Delta \times \frac{2}{7} = 1\frac{1}{2}$

4) $1\frac{3}{8} \times \Delta = 5\frac{1}{2}$

5) $1\frac{1}{2} \times \Delta = 1$

6) $\frac{2}{3} \times \Delta = 1\frac{1}{2}$

**3-28.** Divide the fractions.

$$\frac{1}{2} \div \frac{1}{3}$$

SOLUTION

a) Switch the numerator and denominator of the divisor (the second fraction).
b) Change the operation sign from division to multiplication ($\div \rightarrow \times$).
c) Solve the same way as multiplying fractions.
d) Simplify it.

i) Flip of $\frac{1}{3}$.

$$\frac{1}{2} \div \frac{1}{3} = \frac{1}{2} \times \frac{3}{1} = \frac{3}{2} \text{ or } 1\frac{1}{2} \quad \text{Simplify.}$$

ii) Change the operation.
iii) Then multiply them.

So, the quotient of $\frac{1}{2} \div \frac{1}{3}$ is $1\frac{1}{2}$.

**3-29.** Divide the fractions.

$$1\frac{1}{2} \div \frac{3}{4}$$

SOLUTION

If they are unlike fractions, then i) change the mixed numbers to improper fractions, ii) flip the numerator and denominator of the divisor, iii) change the operation sign, iv) multiply and simplify if necessary

ii) Flip (reciprocal)* of fraction.

i) Change the mixed numbers to improper fractions.

$$1\frac{1}{2} \div \frac{3}{4} \longrightarrow \frac{3}{2} \div \frac{3}{4} = \frac{3}{2} \times \frac{4}{3}$$

iii) Change the operation sign.

iv) Multiply and simplify if necessary.

$$\frac{3}{2} \times \frac{4}{3} \longrightarrow \frac{{}^{1}3}{{}_{1}2} \times \frac{4^{2}}{3_{1}} = 2$$

So, the quotient of $1\frac{1}{2} \div \frac{3}{4}$ is 2.

**Exercises 52**    Divide the fractions and simplify if necessary.

1)    $\dfrac{2}{3} \div \dfrac{3}{4}$

2)    $\dfrac{2}{3} \div \dfrac{4}{9}$

_____

3)    $\dfrac{4}{5} \div \dfrac{3}{8}$

4)    $\dfrac{4}{7} \div \dfrac{3}{4}$

_____

5)    $2\dfrac{2}{5} \div 1\dfrac{4}{5}$

6)    $3\dfrac{1}{8} \div 2\dfrac{1}{2}$

_____

7)    $1\dfrac{1}{3} \div 1\dfrac{1}{6}$

8)    $3\dfrac{1}{2} \div 2\dfrac{1}{6}$

_____

9)    $3\dfrac{1}{7} \div \dfrac{2}{8}$

10)    $\dfrac{2}{3} \div \dfrac{5}{8}$

_____

**Exercises 53**    Find the value using the given information.

1)    If $y = 1\dfrac{1}{2}$, find the value of $y \div 2$.

2)    If $y = 1\dfrac{1}{3}$, find the value of $\dfrac{4}{6} \div y$.

_____

3)    If $y - \dfrac{3}{7} = 2$, find the value of $17 \div y$.

4)    If $1 + y = 1\dfrac{1}{2}$, find the value of $y \div 3\dfrac{1}{2}$.

_____

5)    If $y - 1 = \dfrac{4}{5}$, find the value of $y \div \dfrac{3}{5}$.

6)    If $\dfrac{2}{5} + y = 1$, find the value of $y \div 1\dfrac{1}{2}$.

_____

**Exercises 54**    Divide the fractions and simplify if necessary.

1) $6 \div \dfrac{4}{15}$

2) $3\dfrac{1}{2} \div 1\dfrac{1}{7}$

3) $\dfrac{3}{12} \div \dfrac{6}{10}$

4) $5 \div \dfrac{5}{7}$

5) $2\dfrac{1}{5} \div \dfrac{1}{4}$

6) $2\dfrac{1}{9} \div \dfrac{1}{4}$

7) $2\dfrac{1}{4} \div 1\dfrac{2}{3}$

8) $1\dfrac{1}{2} \div \dfrac{2}{3}$

9) $1\dfrac{1}{3} \div 1\dfrac{1}{5}$

10) $2\dfrac{2}{7} \div 1\dfrac{3}{4}$

11) $2\dfrac{1}{2} \div 1\dfrac{3}{5}$

12) $4\dfrac{1}{8} \div 4$

**Exercises 55**    Find the value of Δ.

1) $\Delta \div \dfrac{2}{7} = 1\dfrac{1}{5}$

2) $\Delta \div \dfrac{1}{3} = 3$

3) $1\dfrac{1}{2} \div \Delta = 1$

4) $\dfrac{3}{5} \div \Delta = \dfrac{1}{2}$

5) $1\dfrac{3}{8} \div \Delta = \dfrac{1}{2}$

6) $\Delta \div \dfrac{5}{12} = \dfrac{1}{3}$

**SELF-TEST**

1.  Which of the following equations are correct?

    **A.** $3 \times \dfrac{1}{5} = \dfrac{1}{5} \times \dfrac{1}{5} \times \dfrac{1}{5}$          **B.** $\dfrac{1}{2} \times 2 = \dfrac{1}{2} \times \dfrac{1}{2}$

    **C.** $4 \times \dfrac{2}{3} = \dfrac{2}{3} + \dfrac{2}{3} + \dfrac{2}{3} + \dfrac{2}{3}$          **D.** $3 \times \dfrac{2}{3} = (3 + 3 + 3) \times \dfrac{2}{3}$

2.  Which of the following equations are correct?

    **A.** $4 \times \dfrac{2}{7} = (4 \times 4 \times 4 \times 4) + \dfrac{2}{7}$          **B.** $3 \times \dfrac{1}{5} = \dfrac{1}{5} \times \dfrac{1}{5} \times \dfrac{1}{5}$

    **C.** $\dfrac{2}{5} \times 3 = \dfrac{2}{5} + \dfrac{2}{5} + \dfrac{2}{5}$          **D.** $\dfrac{1}{2} \times 2 = \dfrac{1}{4}$

3.  Which of the following equations are correct?

    **A.** $3 \div \dfrac{3}{4} = \dfrac{4}{3} \times \dfrac{4}{3} \times \dfrac{4}{3}$          **B.** $3 \div \dfrac{2}{5} = (3 + 3 + 3) \times \dfrac{2}{5}$

    **C.** $\dfrac{1}{3} \div 2 = \dfrac{1}{3} \times \dfrac{1}{3}$          **D.** $\dfrac{1}{6} \div \dfrac{1}{4} = \dfrac{1}{6} + \dfrac{1}{6} + \dfrac{1}{6} + \dfrac{1}{6}$

4.  Which of the following equations are correct?

    **A.** $\dfrac{3}{8} \div 3 = \dfrac{3}{8} + \dfrac{3}{8} + \dfrac{3}{8}$          **B.** $3 \div \dfrac{1}{4} = \dfrac{1}{4} \times \dfrac{1}{4} \times \dfrac{1}{4}$

    **C.** $\dfrac{2}{3} \div 2 = \dfrac{1}{3}$          **D.** $4 \div \dfrac{2}{3} = (4 \times 4 \times 4 \times 4) + \dfrac{3}{2}$

5.  A certain recipe calls for ¼ teaspoons of cayenne pepper, 1/2 teaspoons of black pepper, and 1/3 teaspoons of red pepper. How many teaspoons of pepper are used in the recipe?

    **A.**  2                                        **B.**  $\dfrac{1}{12}$

    **C.**  4                                        **D.**  $1\dfrac{1}{12}$

**6.** Find the quotient.

$$\frac{4}{9} \div \frac{1}{6}$$

**A.** 2

**B.** $\frac{1}{2}$

**C.** 4

**D.** $2\frac{2}{3}$

**7.** What is the value of $\Delta$ for the equation below?

$$1\frac{1}{2} \times \Delta = 4\frac{1}{2}$$

**A.** $1\frac{1}{2}$

**B.** $3\frac{1}{2}$

**C.** 3

**D.** 4

**8.** What is the value of $\Delta$ for the equation below?

$$\Delta \div 2 = \frac{3}{8}$$

**A.** $2\frac{3}{8}$

**B.** $2\frac{3}{4}$

**C.** $\frac{3}{4}$

**D.** $\frac{3}{8}$

**9.** If $x = \frac{2}{5}$, what is the value of $\frac{5}{9} \times (x + \frac{1}{2})$?

**A.** $\frac{1}{2}$

**B.** $\frac{1}{9}$

**C.** $\frac{13}{18}$

**D.** $\frac{5}{21}$

**10.** If $x = 3$, what is the value of $(\frac{3}{4} \div x) + 3$?

**A.** $\frac{1}{8}$

**B.** $5\frac{1}{4}$

**C.** $6\frac{3}{4}$

**D.** $3\frac{1}{4}$

**11.** At a birthday party, the boys ate 2(3/4) pieces of pizza while the girls ate 1(1/2) pieces. How many pieces of pizza were eaten at the party?

A. 5                                        B. $4\frac{1}{4}$

C. 4                                        D. $3\frac{1}{2}$

12. A granola bar recipe calls for 3/8 cups of walnuts and 2/3 cups of almonds. Manuel wants to double the recipe. If so, how many cups of nuts will he need to use?

A. 2                                        B. $\frac{1}{2}$

C. 4                                        D. $2\frac{1}{12}$

13. What is the value of $\Delta$ for the equation below?
$$1 \div \Delta = \frac{5}{7}$$

A. $\frac{5}{7}$                                    B. $1\frac{2}{5}$

C. $1\frac{5}{7}$                                    D. $\frac{2}{5}$

14. If $x + 1 = \frac{1}{2}$, what is the value of $\frac{3}{5} - (1 + x)$?

A. $\frac{1}{10}$                                    B. $\frac{1}{2}$

C. $-\frac{9}{10}$                                   D. $1\frac{3}{5}$

15. If $1 - x = \frac{1}{10}$, what is the value of $1 - (x - 1)$?

A. $-1\frac{1}{10}$                                  B. $-\frac{9}{10}$

C. $2\frac{1}{10}$                                   D. $\frac{1}{10}$

# CHAPTER 4
# Percentage

In this chapter, you will solve problems that involve fractions, decimals, and percentages that can apply to your daily life.

## 1.  Understanding Percent

**4–1.**  Interpret percentage as a fraction of 100.

A percentage is a ratio that is a specified amount in or for every hundred.

**4–2.**  Input 33 shaded squares in a 10 × 10 grid below. Find the fraction and percent of the shaded squares.

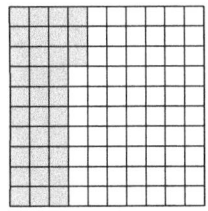

10 x 10 grid

33 shaded squares

**SOLUTION**

a.  33 shaded squares out of 100 squares is $\frac{33}{100}$, which is a fraction.

b.  i) So, you can set up a proportion. Let $x$ represent the variable of an unknown number.

$$\frac{33}{100} = \frac{x}{100\%}$$

ii) Now you can find the value of $x$ by multiplying 100% on both sides.

$$\frac{33}{100} \times 100\% = \frac{x}{\cancel{100\%}} \times \cancel{100\%}$$

$$\frac{33}{100} \times 100\% = x$$

Let $x$ represents the variable.

$$\frac{\cancel{3300}\%^{33}}{\cancel{100}^{1}} = x$$

| **Fraction x 100% = $x$** |
| --- |

\* Remembers the equation to turn a fraction into a percentage.

$$33\% = x$$

So, 33 shaded squares out of 100 is 33%.

133

**4-3.**   Write $\frac{4}{5}$ as a percentage.

> **SOLUTION**
>
> You can apply the equation to convert the fraction as a percentage.
> Let $x$ represents the variable of an unknown percentage.
> Fraction $\times$ 100% = $x$    Apply as an equation.
>
> $\frac{4}{5} \times 100\% = x$                       Substitute "Fraction" with $\frac{4}{5}$.
>
> $\frac{4 \times 100\%}{5} = x$                        Simplify.
>
> $\frac{400\%}{5} = x$                              Divide.
>
> $80\% = x$
>
> So, $\frac{4}{5}$ equals to 80%.

**Exercises 1**    Find the fraction and percentage that represents the shaded segments.

**1)**
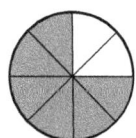

Fraction =
Percent (%) =

**2)**
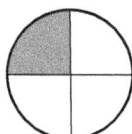

Fraction =
Percent (%) =

**3)**

Fraction =
Percent (%) =

**4)**

Fraction =
Percent (%) =

**5)**
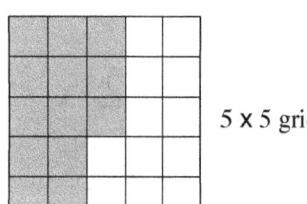      5 x 5 grid

Fraction =
Percent (%) =

**6)**
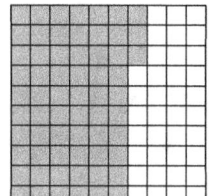      10 x 10 grid

Fraction =
Percent (%) =

**7)**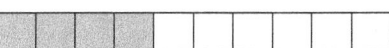

Fraction =
Percent (%) =

**8)**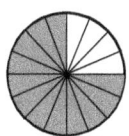

Fraction =
Percent (%) =

**Exercises 2**    Write each fraction as a percentage.

**1)**  $\dfrac{1}{50}$                                    **2)**  $\dfrac{3}{20}$

_____

**3)**  $\dfrac{6}{10}$                                    **4)**  $3\dfrac{40}{50}$

_____

**5)**  $2\dfrac{3}{4}$                                    **6)**  $\dfrac{5}{100}$

_____

**7)**  $1\dfrac{1}{5}$                                    **8)**  $\dfrac{28}{100}$

_____

**9)**  $\dfrac{8}{40}$                                    **10)**  $1\dfrac{1}{2}$

_____

**Exercises 3**    Write each percentage as a fraction.

**1)**  3%                                    **2)**  130%

_____

**3)**  19%                                    **4)**  25.18%

_____

**5)**  0.08%                                    **6)**  3.7%

_____

**7)**  0.95%                                    **8)**  92%

_____

**9)**  250%                                    **10)**  99.99%

_____

# * Solving Problems

**Exercises 4**   Melisa's class is participating in a fire drill. There are 29 students total in her class. Two of her classmates were absent today.

**1)** What is the fraction of the students who attended the drill? Explain your answer.

**2)** What percentage of Melisa's classmates is participating in the fire drill? Explain your answer.

**3)** Chris is measuring how much water he has. There is 684 mL left from a 3800 mL bottle.
What is the percentage of the water he drank? Explain your answer.

**Exercises 5**   There are 15 red balloons, 19 blue balloons, and 8 yellow balloons at a birthday party.

**1)** What is the percentage of blue balloons? Explain your answer.

**2)** What is the percentage of yellow balloons? Explain your answer.

**1.** What is the percentage of the shaded area?

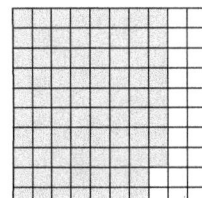

      **A.**  50%         **B.**  100%

      **C.**  120%     **D.**  150%

\* Use the figure for Exercises **2-3**.

**2.** What is the fraction of the shaded area?

    **A.** $\dfrac{78}{100}$      **B.** $\dfrac{22}{78}$

    **C.** $\dfrac{78}{22}$      **D.** $\dfrac{100}{78}$

**3.** What is the percentage of the shaded area?

    **A.**  23%             **B.**  64%

    **C.**  78%             **D.**  96%

**4.** Which of the following is $\dfrac{7}{5}$ written as a percentage? Round to the nearest tenth.

    **A.**  1.4%          **B.**  71.4%

    **C.**  140%         **D.**  150%

**5.** Bess has $12.00. She wants to buy a book that cost $9.00. Find the percentage of the money she would have to spend.

    **A.**  75%          **B.**  15%

    **C.**  25%          **D.**  85%

**6.** Camry bought 72 pieces of candy for everyone in her class. If she gave 4 pieces to everyone, what is the percentage of candy that each student received? Round to the nearest tenth.

    **A.**  180%         **B.**  5.6%

    **C.**  94.4%       **D.**  105%

7. Which of the following is $1\frac{2}{5}$ written as a percentage? Round to the nearest whole number.

   A. 200%                                    B. 140%
   C. 20%                                     D. 40%

8. At a potluck, 74 out of 102 hot dogs are eaten. What is the percentage of hot dogs left? Round to the nearest tenth.

   A. 72.5%                                   B. 137.8%
   C. 27.5%                                   D. 37.8%

9. Iggy, Eunice, and Lily each blow up 28 balloons. If there are 100 balloons available, what is the percentage of balloons left? Round to the nearest whole number.

   A. 28%                                     B. 16%
   C. 32%                                     D. 84%

10. Which of the following is $\frac{1}{6}$ written as a percent? Round to the nearest whole number.

   A. 10%                                     B. 60%
   C. 17%                                     D. 34%

11. There are 163 pigeons in the park. After a while, 78 of them fly away. What is the percentage of pigeons left in the park? Round to the nearest whole number.

   A. 48%                                     B. 50%
   C. 52%                                     D. 54%

12. Chris received $45.00 from his mother and $35.00 from his father. What is the percentage of the money he received from his mother? Round to the nearest whole number.

   A. 78%                                     B. 22%
   C. 129%                                    D. 44%

## 2.  Change Decimals as a Percent

**4-4.** Interpret percentage as a decimal

> **SOLUTION**
>
> Use the equation to turn a decimal into a percentage. It is done similarly with fractions.
>
> Let $x$ represents the variable of an unknown percentage.
>
> $$\boxed{\textbf{Decimal} \times \textbf{100\%} = x}$$

*Remember the equation to turn a decimal into a percentage

**4-5.** What is 0.78 written as a percentage?

> **SOLUTION**
>
> Use the equation above.
> | | |
> |---|---|
> | Decimal $\times$ 100% = $x$ | Apply the equation. |
> | $0.78 \times 100\% = x$ | Substitute "Decimal" with 0.78. |
> | $78\% = x$ | Simplify. |
>
> So a percent (%) of 0.78 is 78%.

**Exercises 6**    Write each decimal as a percentage.

1)  0.01

2)  0.36

3)  0.5

4)  1.02

5)  0.82

6)  1.46

7)  0.70

8)  0.591

9)  1.078

10)  0.3064

11) 3.526

12) 9.01

**Exercises 7**   Write each fraction or decimal as a percentage.

1)  $\frac{5}{8}$

2)  0.63

_____                    _____

3)  $\frac{15}{1000}$

4)  0.002

_____                    _____

5)  1.050

6)  $\frac{12}{10}$

_____                    _____

7)  $1\frac{1}{8}$

8)  12.200

_____                    _____


**Exercises 8**   Write each percentage as a decimal.

1)  2%

2)  105%

_____                    _____

3)  47%

4)  92%

_____                    _____

5)  32%

6)  120%

_____                    _____


**Exercises 9**   Write each percentage as a fraction.

1)  9%

2)  10.5%

_____                    _____

3)  12%

4)  250%

_____                    _____

5)  98%

6)  1.5%

_____                    _____

## * Solving Problems

**Exercises 10**    0.9015 inches of a nail is sticking outside a wall. If the full size of a nail is 2.273 inches long, find the percentage of the length of the nail sticking outside the wall.

1. Which of the following is 0.36 written as a percentage?

    **A.** 36%                                    **B.** 64%
    **C.** 79%                                    **D.** 96%

2. Which of the following is 0.902 written as a percentage?

    **A.** 10.8%                                 **B.** 100%
    **C.** 9.0%                                  **D.** 90.2%

* Use the information below for Exercises **3-4**.
   Mark has 4 pool sets and 60 pool balls. Each pool set has the same number of balls
3. How many balls are there in a set?

    **A.** 19                                    **B.** 15
    **C.** 45                                    **D.** 60

4. What is the percentage of the number of pool balls in a set out of the total number of balls?

    **A.** 25%                                   **B.** 27%
    **C.** 75%                                   **D.** 73%

* Use the information below for Exercises **5-7**.
5. A squirrel buries the same amount of acorns each day. If the squirrel buries 98 acorns in a week, how many acorns did it bury each day?

    **A.** 7                                     **B.** 14
    **C.** 91                                    **D.** 686

6. What is the percentage of acorns buried each day?

**A.** 14%                                    **B.** 7%
**C.** 93%                                    **D.** 86%

7.  What is the percentage of acorns buried for five days?

    **A.** 29%                                **B.** 67%
    **C.** 71%                                **D.** 85%

8.  Which of the following is 1.04 written as a percentage?

    **A.** 4%                                 **B.** 1%
    **C.** 96%                                **D.** 104%

9.  Jake is painting a house. It will take 56 hours before the house is finished. So far, Jake has painted for 29 hours. How much longer will it take for him to finish painting the house? Write your answer as a percentage. Round to the nearest whole number.

    **A.** 48%                                **B.** 34%
    **C.** 52%                                **D.** 66%

10. 7 dresses require 56 feet of fabric to make. If there is 144 feet of fabric available, how many percent of fabric can be left?

    **A.** 39%                                **B.** 12.5%
    **C.** 87.5%                              **D.** 61%

11. Which of the following is 1.101 written as a percentage?

    **A.** 1.1%                               **B.** 101%
    **C.** 110.1%                             **D.** 220%

12. Karl scored 87 points in his science test. What is his score if the test was out of 120 points?

    **A.** 27.5%                              **B.** 44%
    **C.** 72.5%                              **D.** 87%

13. Which of the following is 21.21 written as a percentage?

    **A.** 2,121%                             **B.** 21%
    **C.** 79%                                **D.** 212%

## 3.  Solving Problems with Percentages

**4–6.**  What is 25% of 90?

SOLUTION

You can solve this kind of problems by using multiplying the given numbers. Let $x$ represents the variable of an unknown number.

What is 25% of 90?

$$\frac{1}{100\%} \times \text{Given Percent (\%)} \times \text{Given Number} = x$$

Use the equation above and substitute in the known values.

$$\frac{1}{100\%} \times 25\% \times 90 = x \qquad \text{Substitute.}$$

You can cancel the units (%) using the GCF and simplify them before multiplying.

$$\frac{1}{\cancel{100\%}_{4}} \times \cancel{25\%}^{1} \times 90 = x \qquad \text{Simplify 25 and 100 using the GCF of 25.}$$

$1 \times 25 = 25$
$4 \times 25 = 100$

$$\frac{90}{4} = x \qquad \text{Simplify.}$$

$$22.5 = x \qquad \text{Simplify.}$$

So, 25% of 90 is 22.5.

**Exercises 11**   Write as a decimal and a fraction.

1)   0.1%

2)   0.01%

3)   0.101%

4)   1.01%

5)   0.293%

6)   3.04%

7)   10.81%

8)   3.25%

9)   0.305%

10)   0.064%

**Exercises 12**    Solve each expression.

1)  What is 0.1% of 25?                    2)  What is 0.75% of 15?

_____                            _____

3)  What is 300% of 9?                     4)  What is 1.8% of 78?

_____                            _____

5)  What is 0.01% of 10?                   6)  What is 0.102% of 100?

_____                            _____

7)  What is 3% of 20?                      8)  What is 1.00% of 198?

_____                            _____

9)  What is 128% of $30.00?                10)  What is 12% of $45.00?

_____                            _____

11)  What is 0.3% of 200?                  12)  What is 9% of 20?

_____                            _____


**4–7.**    Complete the sentence below.

38% of 25 is (     ).

**SOLUTION**

Use the same equation and let $x$ represent the unknown number.

38% of 25 is (          ).

$$\frac{1}{100\%} \times \textbf{Given Percent (\%)} \times \textbf{Given Number} = x$$

$$\frac{1}{100\%} \times 38\% \times 25 = x \qquad \text{Substitute.}$$

You can cancel the units (%) and simplify.

$$\frac{1}{\underset{50}{\cancel{100\%}}} \times \overset{19}{\cancel{38\%}} \times 25 = x \qquad \text{Simplify 100 and 38 using the GCF of 2.}$$

Simplify again until you find the lowest prime numbers.

$$\frac{1}{\underset{\cancel{50}\,2}{\cancel{100\%}}} \times \overset{19}{\cancel{38\%}} \times \underset{1}{\cancel{25}} = x \qquad \text{Simplify 50 and 25 using the GCF of 25.}$$

$$\frac{19}{2} = x \qquad\qquad\qquad \text{Simplify.}$$

$$9.5 = x \qquad\qquad\qquad \text{Simplify.}$$

So, 38% of 25 is 9.5.

**Exercises 13**    Complete each statement.

**1)**  1.5% of \$4 is  (              ).

**2)**  0.85% of 20 is (      ).

_____

**3)**  120% of \$60 is (          ).

**4)**  5.5% of 320 is (      ).

_____

**5)**  0.02% of 100 is (        ).

**6)**  0.8% of 25 is (      ).

_____

**7)**  15% of 85 is (        ).

**8)**  220% of 50 is (        ).

_____

**9)**  65% of \$35.00 is (        ).

**10)**  1% of \$250.00 is (        ).

_____

**11)**  8.5% of \$82 is (        ).

**12)** 92% of 125 is (        ).

_____

**13)**  3% of \$40 is (        ).

**14)** 12% of 20 is (        ).

_____

**Exercises 14**   Solve each expression.

1)  1% of 50

2)  20% of 30

3)  92% of 24

4)  0.5% of 50

5)  200% of 60

6)  100% of 20

7)  12% of $75

8)  38% of 8

9)  2% of 580

10)  95% of 250

**Exercises 15**   Solve each expression.

1)  95% of 120 is (        ).

2)  What is 2.0% of $120?

3)  What is 0.100% of $120?

4)  0.75% of $15 is (        ).

5)  1.5% of $4 is  (        ).

6)  What is 4% of $18?

7)  150% of $60 is (        ).

8)  0.8% of $180 is (        ).

9)  What is 40% of 55?

10)  What is 0.1% of $98?

# * Solving Problems

**Exercises 16**    Solve each problem using the given information.

    **1)** A whole pizza weighs 3.8 lbs. If the pizza is exactly divided into 8 slices, how many pounds does slice R weigh?

    **2)** What is the percentage of the weight of each slice?

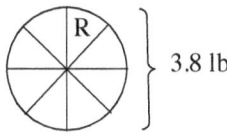

**Exercises 17**    There are 35 apples total in 5 boxes. Each box represents a certain number of apples. What is the percentage of the number of apples in a box?

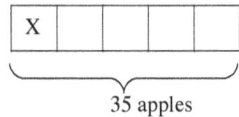

**Exercises 18**    Solve each problem using the given information.

| 1) Shade 125% on the figure below. 2) If you shade in 13 slices, what is the percentage of the shaded slices? | 3) Shade 24% on the 5 by 5 grid. 4) If you shade in 12 boxes, what is the percentage of the shaded boxes? |
|---|---|
| 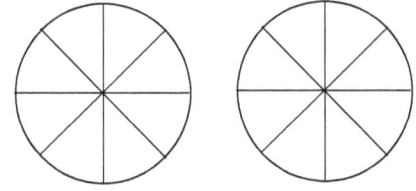 | 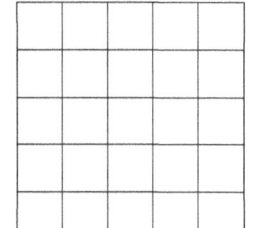 |

1. What is 10% of 18?

   **A.** 1.8                                 **B.** 18
   **C.** 16.2                                **D.** 180

2. Complete the sentence.
   64% of 90 is (          ).

   **A.** 57.6                                **B.** 140.6
   **C.** 100                                 **D.** 5,760

3. Jonathan wants to buy a game that costs $35.00. The sales tax is 9%. How much money does he need to pay if he wants to buy two copies?

   **A.** $38.15                              **B.** $133.00
   **C.** $76.30                              **D.** $114.45

4. What is 95% of 8?

   **A.** 7.6                                 **B.** 8
   **C.** 8.4                                 **D.** 8.6

5. Joan is in the store to buy a dress that costs $45.00. All items are 40% off today. How much money does she pay for the dress, given she does not have to pay tax?

   **A.** $18.00                              **B.** $27.00
   **C.** $72.00                              **D.** $112.5.00

6. What is 72% of 49?

   **A.** 35.28                               **B.** 13.72
   **C.** 68.06                               **D.** 175

7. If a container can hold up to 50 bottles of water, how many bottles of water can fill 50% of the container?

   **A.** 25                                  **B.** 100
   **C.** 50                                  **D.** 250

**8.** If a pizza has 16 slices, what is 75% of the pizza?

      **A.**  12                            **B.**  16
      **C.**  18                            **D.**  28

**9.** Complete the sentence.
      8% of $15 is  (       ).

      **A.**  1.2                         **B.**  12
      **C.**  1.8                         **D.**  187.5

**10.** Jess paid $4.30 for a juice in a store. What is the original price for the juice if the sales tax is 8%?

      **A.**  $0.34                    **B.**  $4.64
      **C.**  $4.25                    **D.**  $3.96

**11.** What is 125% of 125?

      **A.**  156.25                  **B.**  1
      **C.**  100                      **D.**  15.6

**12.** Complete the sentence.
      13.5% of $25 is  (        ).

      **A.**  3.38                      **B.**  337.5
      **C.**  0.19                      **D.**  185.2

# CHAPTER 5
# Measurements and Geometry

In this chapter, you will learn about the different properties of perpendicular and parallel lines, and identify distinctive pairs of angle relationships. You will also be finding the area, perimeter, and volume of various shapes and solids.

## 1. Understanding Plane

**5-1.** Measurements

| Vocabulary | Definations/Characteristics | Diagram | Symbols |
|---|---|---|---|
| Point | No dimension and represented by a small dot. | • A | point A |
| Line | No endpoints and extends forever in two directions. | •A  •B | $\overleftrightarrow{AB}$ |
| Line segment | A part of a line that is bounded by two distict end points. | •A  •B | $\overline{AB}$ |
| Ray | One endpoint and extends forever in one direction. | •A  •B | $\overrightarrow{AB}$ |
| Plane | Extends forever in all directions. | A •C B | plane AB |

Collinear: On the same line.        Coplanar: On the same plane.

**5-2.** Congruent, Similar, and Transformations

1) **Congruent:** If figures are congruent, that means they have the same shape, size, and angle.

congruent

2) **Similar:** Figures that have the same shape, but different sizes.

similar

150

**3)** Transformations: Figures in a plane can move in different ways by translating, reflecting, or rotating to produce new figures.

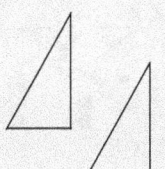

translation; slide

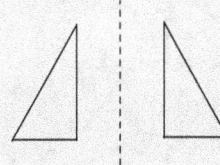

reflection; flip

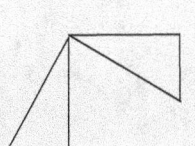

rotation; a point

**Exercises 1**    Use the diagram at the right. Name each figure.

1)  $\overrightarrow{AB}$

2)  $\overline{AC}$

3)  $\overleftrightarrow{BD}$

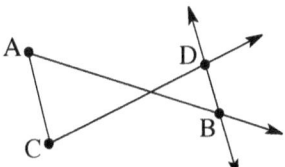

**Exercises 2**    Determine the relationship of each pair.

1)

2)

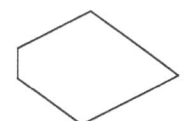

**Exercises 3**    Match each term to the correct figure.
   A. Parallel lines        B. Perpendicular lines        C. Intersecting lines
   D. Acute angle           E. Right angle                F. Obtuse angle

1)

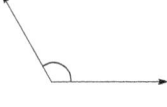

_____

2)

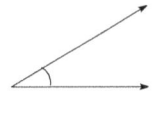

_____

3)

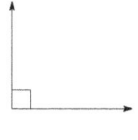

_____

4)

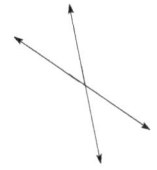

_____

5)

_____

6)

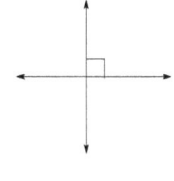

_____

**Exercises 4**    Determine how ABC moves to become EFG.

1)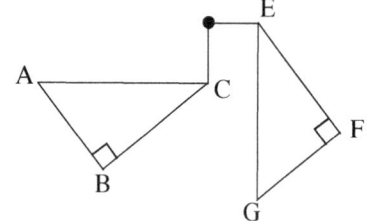

2)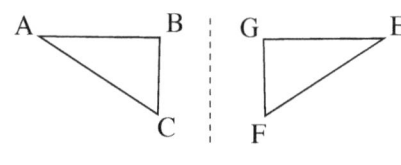

3)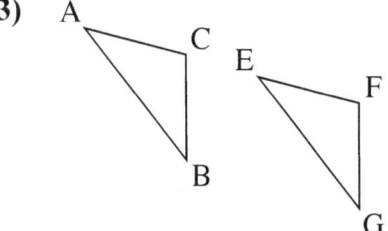

4)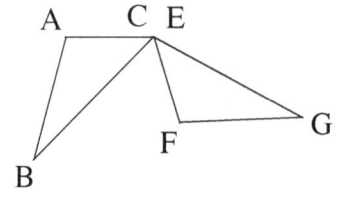

**Exercises 5**    Match each term to its correct definition.

A. Right angle              B. Acute angle              C. Obtuse angle
D. Perpendicular lines      E. Parallel lines           F. Intersecting lines

**1)** An angle whose measure is less than 90°.

**2)** An angle that is 90°.

**3)** An angle that is between 90° and 180°.

**4)** Two lines that forms a right angle.

**5)** Two lines that will never intersect.

**6)** There are two lines that intersect but do not form a right angle

**SELF-TEST**

1. Which of the following terms best describes two or more shapes that are exactly the same?

   **A.** Plane                                    **B.** Congruent
   **C.** Ray                                      **D.** Line segment

2. Which of the following terms best describes points that lie on the same line?

   **A.** Collinear points                         **B.** Ray
   **C.** Coplanar points                          **D.** Plane

3. What is an angle whose measure is between $90^\circ$ and $180^\circ$?

   **A.** Obtuse angle                             **B.** Acute angle
   **C.** Right angle                              **D.** Intersecting lines

4. Which of the following figures are perpendicular lines?

   **A.**                                          **B.**

   **C.**                                          **D.**

5. If $\angle DAC = \angle BAD$, which of the following terms best describes this statement?

   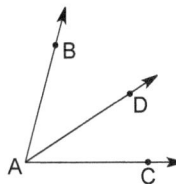

   **A.** Segment Addition Property
   **B.** Angle Addition Property
   **C.** Congruent
   **D.** Ray

6. Two triangles have the same shape, but are different sizes. What is the best term for this statement?

   **A.** Congruent                                **B.** Similar
   **C.** Translation                              **D.** Transformation

7. What is the best way to describe translation?

    **A.** A figure in a plane rotating in different directions.
    **B.** A figure in a plane sliding in different directions.
    **C.** A figure in a plane flipping into its mirror image.
    **D.** A figure in a plane moving in different ways.

8. Two figures have the same shape, size, and angle measures. Which of the following terms best describes this statement?

    **A.** Congruent              **B.** Similar
    **C.** Translation           **D.** Transformation

9. Which of the following figures is an acute angle?

    **A.**                                    **B.**

    **C.**                                    **D.**

10. What kind of transformation is depicted in the graph below?

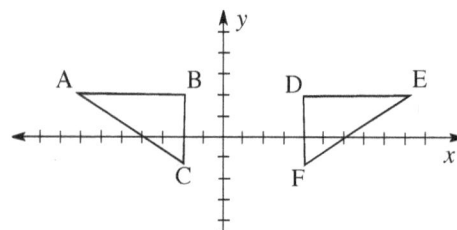

    **A.** Rotation
    **B.** Reflection
    **C.** Translation
    **D.** Slide

11. What kind of transformation is depicted in the graph below?

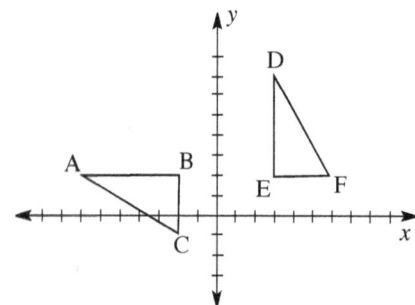

    **A.** Flip
    **B.** Rotation
    **C.** Translation
    **D.** Slide

## 2. Perimeters and Areas of Polygon

**5–3.** A polygon ("poly" meaning many) is a closed plane figure with three or more sides and angles and no curves.

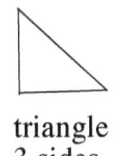
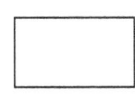
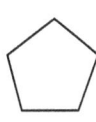

triangle      quadrilateral      pentagon      hexagon      heptagon
3 sides       4 sides            5 sides       6 sides      7 sides

**Exercises 6**    Name each figure that appears below. Circle the ones that are solids.

1)
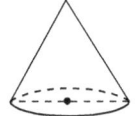

2)
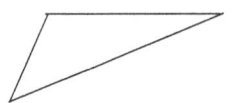

3)
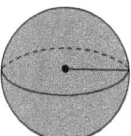

4)
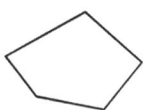

5)

6)
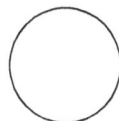

**5–4.**    Area of a polygon

| a)  Rectangle | b)  Parallelogram | c)  Triangle |
|---|---|---|
| 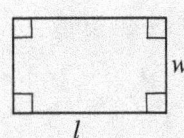 | 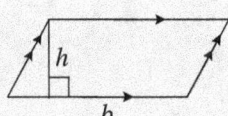 | 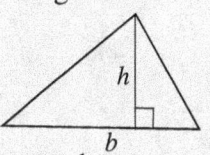 |
| Area = length x width<br>A = $lw$ | Area = base x height<br>A = $bh$ | Area = $\frac{1}{2}$(base x height)<br>A = $\frac{1}{2}bh$ |

**5–5.**    Perimeter (P) of a polygon

In order to find the perimeter of a polygon, add the length of each side.

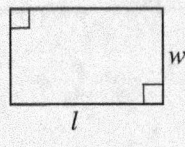
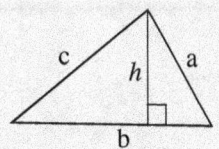
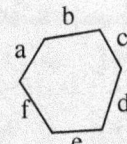

P = 2($l$ + $w$)          P = a + b + c          P = a + b + c + d + e + f

**Exercises 7**    Find the perimeter of each figure.

1)
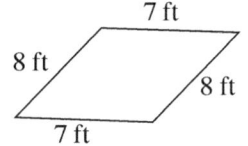

2)
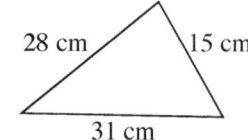

3)
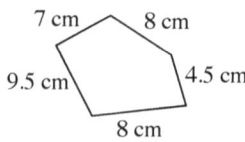

**Exercises 8**    Find the perimeter and area of each figure.

1)
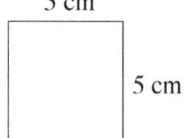

2)
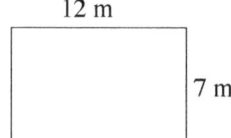

3)
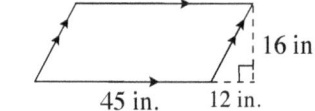

4)
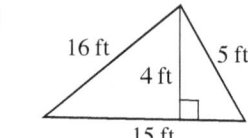

**Exercises 9**    Find each length using the given information in the diagram.

1)    Find the length of the side.

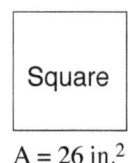

2)    Find the length of *b*.

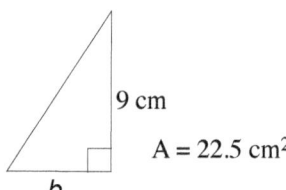

3)    Find the height *h*.

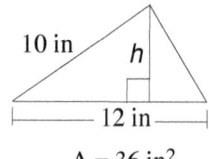

4)    Find the length of *w*.

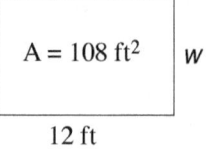

**SELF-TEST**

1.  What is the name of a polygon that has 5 sides?

    **A.** triangle                        **B.** square
    **C.** pentagon                        **D.** parallelogram

2.  What is the perimeter for the rectangle below?

    10 yd          **A.** 56 yd$^2$          **B.** 56 yd
                   **C.** 180 yd            **D.** 180 yd$^2$
    18 yd

3.  Both triangles below have a perimeter of 12 ft. What is the length of the unknown sides?

    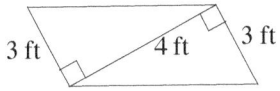
    3 ft   4 ft   3 ft        **A.** 3 ft            **B.** 4 ft
                              **C.** 6 ft            **D.** 300 ft

4.  Which of the following triangles can be classified as isosceles?

    **A.** A triangle that has 8 ft, 9 ft, and 11 ft long sides.
    **B.** A triangle that has at least two congruent sides.
    **C.** A triangle that has at least one right angle.
    **D.** A triangle that has no congruent sides and angles.

5.  What is the perimeter of the figure below?

    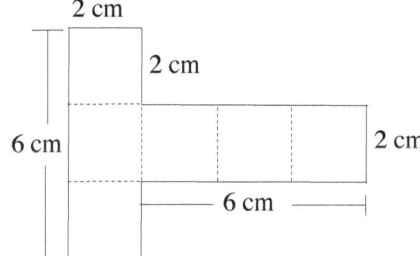
    2 cm
    2 cm            **A.** 28 cm$^2$          **B.** 28 cm
    6 cm     2 cm   **C.** 30 cm             **D.** 30 cm$^2$
         6 cm

## 3. Surface Area of Prisms

**5–6.** Surface area of a rectangular prism:

Construct a cube and a rectangular box from the two-dimensional patterns below and use these patterns to compute the surface area for these objects

A cube has 6 faces, 12 edges, and 8 vertices.
You can find the surface area of a rectangle prism by finding the sum of the area of the faces.

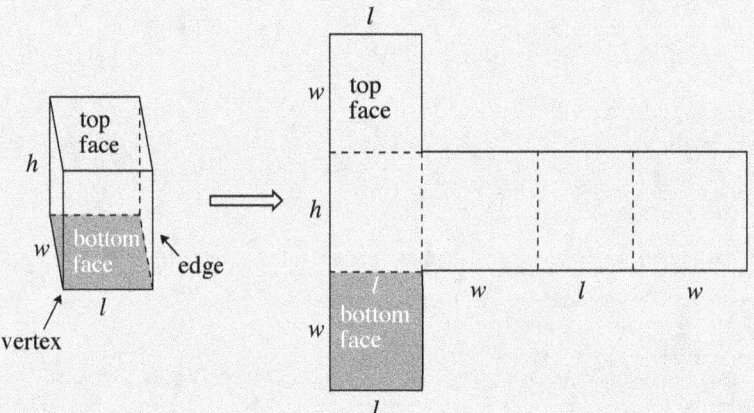

$$\text{surface area}_{\text{rectangle}} = [2(l + w) \times h] + 2(l \times w)$$

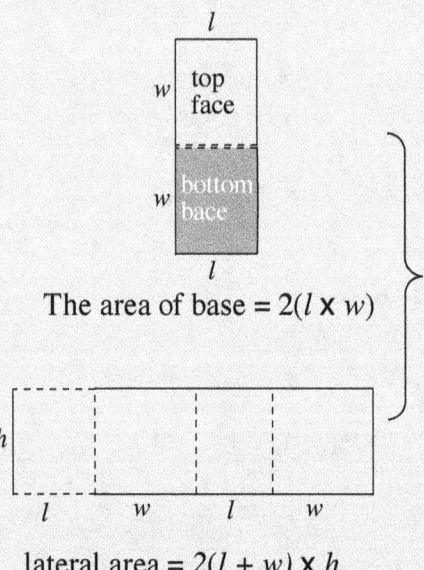

The area of base = $2(l \times w)$

lateral area = $2(l + w) \times h$

The surface area of a prism is the sum of the area of its lateral faces and the base area.

lateral area = $2(l + w)h$, where $l$ is the length and $w$ is the width of the base and $h$ is the height.

The area of bases has two faces of top and bottom parts of a prism.
The area of base = $2 (l \times w)$, where $l$ is the length and $w$ is the width of each face.

Surface area = base area + lateral area
$$= [2(l + w) \times h] + 2(l \times w)$$

**5–7.** Three-dimensional polygons

| Name of solid | Shape | Number of edges | Number of vertices | Number of faces |
|---|---|---|---|---|
| Triangular pyramid | | 6 | 4 | 4 |
| Square pyramid | | 8 | 5 | 5 |
| Cube | | 12 | 8 | 6 |
| Triangular prism | | 9 | 6 | 5 |
| Rectangular prism | | 12 | 8 | 6 |
| Cone | | no | no | no |
| Cylinder | | no | no | no |
| Sphere | | no | no | no |

## 4.  Volume of Prism

**5–8.** Volume of a prism

The volume of a rectangular prism is the product of its length, width, and height.

$V = l \times w \times h$, where $l$ is the length, $w$ is the width, and $h$ is the height of the rectagular prism.

**5-9.** Find the volume of the box.

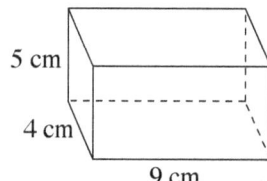

5 cm

4 cm

9 cm

**SOLUTION**

Use the formula of the volume of the rectangular prism. $V = l \times w \times h$, where $l$ is the length, $w$ is the width, and $h$ is the height of the box. $V = 9$ cm x 4 cm x 5 cm $= 180$ cm$^3$

So the volume of the box is 180 cm$^3$. Remember that when writing the volume, put it in cubic units or units$^3$.

**Exercises 10**    Name each solid.

A.  Square pyramid        B.  Sphere              C.  Cone
D.  Triangular prism      E.  Hexangular prism    F.  Cube
G.  Rectangular prism     H.  Cylinder            I.  Triangular pyramid

**1)**

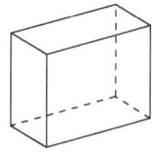

**2)**

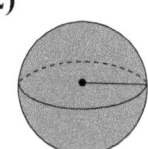

**3)**

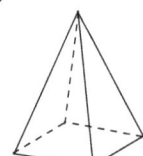

**4)**

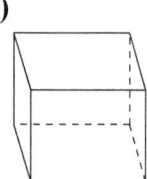

**5)**

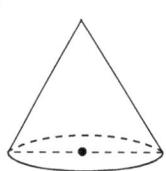

**6)**

**7)**

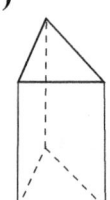

**8)**

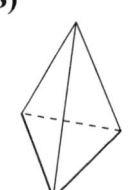

**9)**

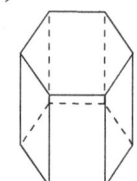

**Exercises 11**    Find the total number of faces, vertices, and edges of each solid.

**1)**                    **2)**                    **3)**

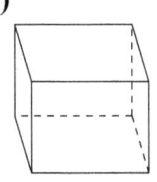    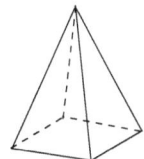    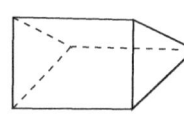

faces: _____          faces: _____          faces: _____
vertices: _____         vertices: _____         vertices: _____
edges: _____          edges: _____          edges: _____

**Exercises 12**    Use the diagram below. What is the solid that it folds into?

**1)**                    **2)**                    **3)**

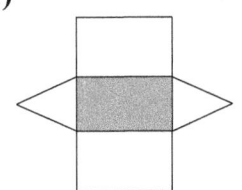    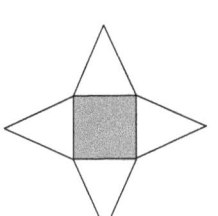    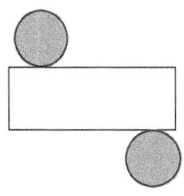

**Exercises 13**    Find the surface area of the figure below.

**1)**

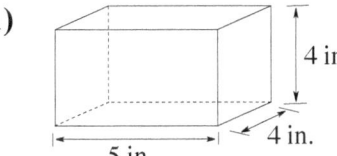

4 in.

4 in.

5 in.

**2)**

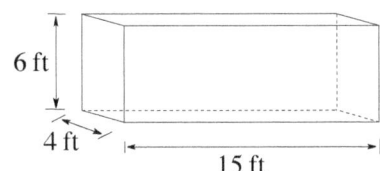

6 ft

4 ft

15 ft

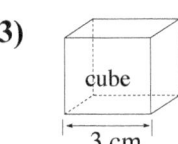

**3)**

cube

3 cm

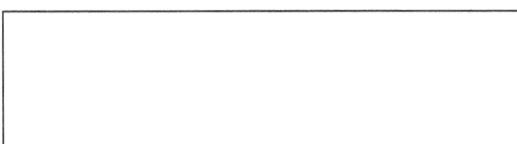

**Exercises 14**    Find the surface area of the figure below.

**1)**

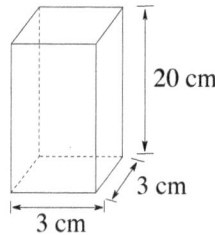

20 cm

3 cm

3 cm

**2)**

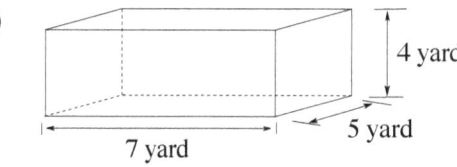

4 yard

7 yard

5 yard

**3)**

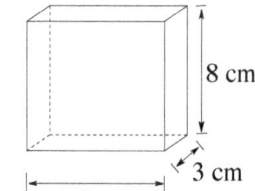

8 cm

3 cm

9 cm

**SELF-TEST**

**1.** What is the surface area of the rectangular prism below?

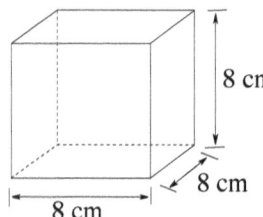

8 cm

8 cm

8 cm

**A.** 384 cm$^2$          **B.** 128 cm$^2$
**C.** 256 cm$^2$          **D.** 512 cm$^2$

**2.** What is the lateral area of the rectangular prism below?

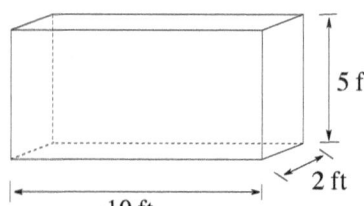

5 ft

10 ft

2 ft

**A.** 20 ft$^2$          **B.** 40 ft$^2$
**C.** 80 ft$^2$          **D.** 120 ft$^2$

**3.** What is the surface area of the rectangular prism below?

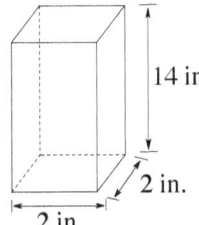

14 in.

2 in.

2 in.

**A.** 56 in$^2$          **B.** 18 in$^2$
**C.** 112 in$^2$         **D.** 36 in$^2$

**4.** What is the volume of the rectangular prism below?

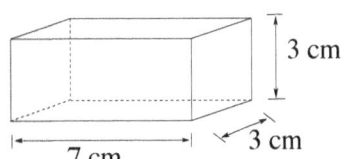

3 cm

3 cm

7 cm

**A.** 21 cm$^3$          **B.** 63 cm$^3$
**C.** 9 cm$^3$           **D.** 30 cm$^3$

**5.** Find the ratio of the volumes of the two figures below.

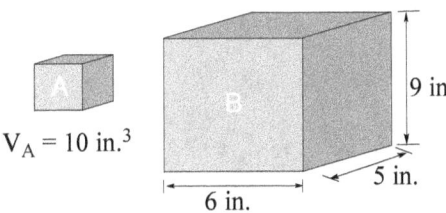

$V_A = 10$ in.$^3$

9 in.

5 in.

6 in.

**A.** 1:2          **B.** 1:10
**C.** 1:15         **D.** 1:27

**6.** What is the volume of the rectangular prism below?

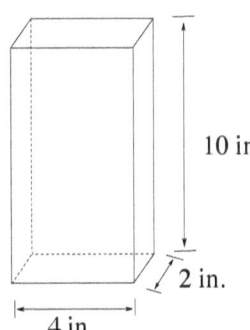

10 in.

2 in.

4 in.

**A.** 8 in$^3$          **B.** 20 in$^3$
**C.** 40 in$^3$         **D.** 80 in$^3$

**7.** Find the ratio of the volumes of the two figures below.

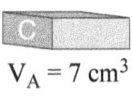

$V_A = 7$ cm$^3$

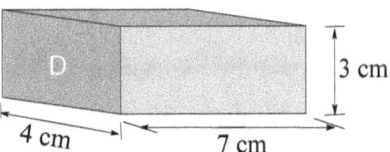

3 cm

4 cm          7 cm

**A.** 1:3          **B.** 1:4
**C.** 1:10         **D.** 1:12

**8.** Find the ratio of the volumes of the two figures below.

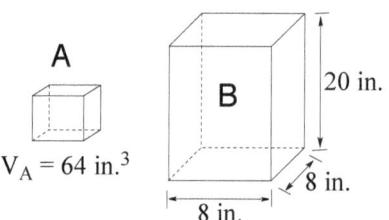

**A.** 1:20          **B.** 1:15
**C.** 1:10          **D.** 1:5

**9.** What is the volume of the rectangular prism below?

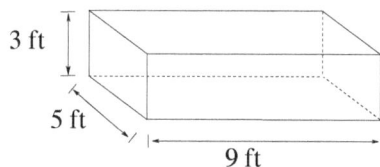

**A.** 45 ft$^3$          **B.** 135 ft$^3$
**C.** 27 ft$^3$          **D.** 84 ft$^3$

**10.** How many edges does a rectangular pyramid have?

**A.** 4                    **B.** 6
**C.** 8                    **D.** 10

**11.** How many faces does a rectangular pyramid have?

**A.** 5                    **B.** 6
**C.** 7                    **D.** 8

**12.** There are 12 edges, 8 vertices, and 6 faces of a three-dimensional polygon. What shape is being described?

**A.** Rectangular prism          **B.** Triangular pyramid
**C.** Cylinder                          **D.** Sphere

**13.** If a cardboard box has a volume of 120 cm$^3$, a width of 4 cm, and a height of 5 cm, what is its length?

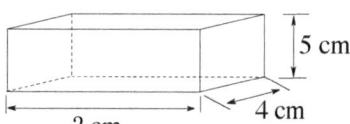

**A.** 3 cm          **B.** 4 cm
**C.** 5 cm          **D.** 6 cm

**14.** A cardboard box has a surface area of 190 yds$^2$, a length of 7 yd, and a width of 2 yd.. What is the height?

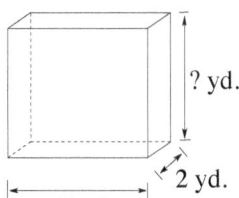

A.  12 yd          B.  9 yd
C.  65 yd          D.  18 yd

15.  The surface areas of **A** and **B** are 108 in$^2$ and 4800 in$^2$ respectively.  What is the volume of **A**?

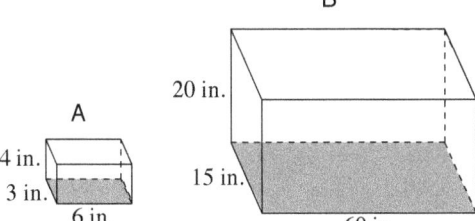

A.  12 in$^3$          B.  18 in$^3$
C.  24 in$^3$          D.  72 in$^3$

16.  Use the diagram and the given information from Question **15**.  Which of the following is the ratio of **A** and **B**?

A.  1:250          B.  1:150
C.  1:125          D.  1:75

17.  Which of the following is the correct unit for the volume?

A.  unit          B.  units$^3$
C.  units$^2$          D.  units$^4$

18.  Which of the following has no faces?

A.  Rectangular prism          B.  Triangular pyramid
C.  Cube          D.  Sphere

19.  Which of the following has no vertices?

A.  Cylinder          B.  Triangular pyramid
C.  Cube          D.  Rectangular prism

20.  What solid shape does the figure below make?

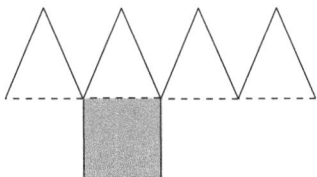

A.  Rectangular prism
B.  Triangular pyramid
C.  Cube
D.  Square pyramid

## 5.  Classifying Triangles

**5–10.**  Classify the triangle.

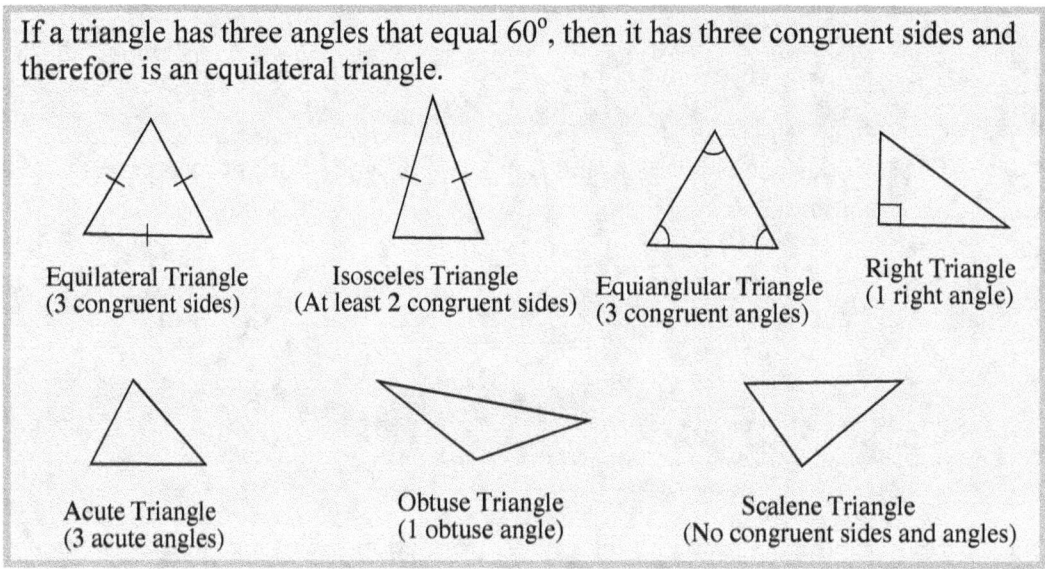

If a triangle has three angles that equal 60°, then it has three congruent sides and therefore is an equilateral triangle.

Equilateral Triangle (3 congruent sides)

Isosceles Triangle (At least 2 congruent sides)

Equianglular Triangle (3 congruent angles)

Right Triangle (1 right angle)

Acute Triangle (3 acute angles)

Obtuse Triangle (1 obtuse angle)

Scalene Triangle (No congruent sides and angles)

**5–11.**   Triangles

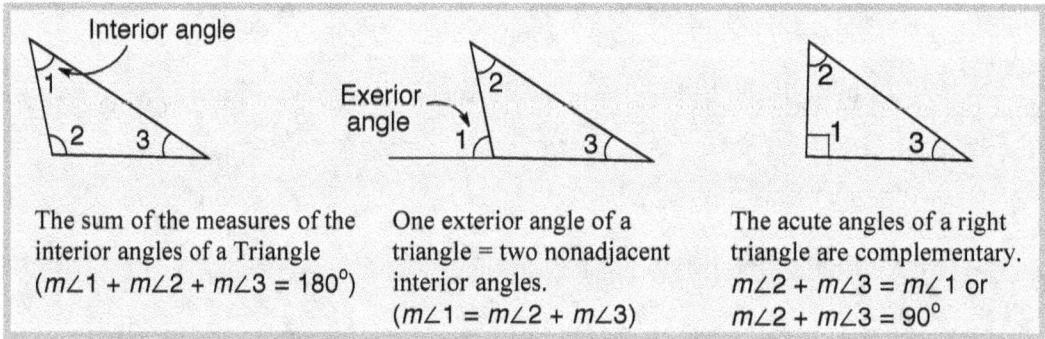

The sum of the measures of the interior angles of a Triangle ($m\angle 1 + m\angle 2 + m\angle 3 = 180°$)

One exterior angle of a triangle = two nonadjacent interior angles. ($m\angle 1 = m\angle 2 + m\angle 3$)

The acute angles of a right triangle are complementary. $m\angle 2 + m\angle 3 = m\angle 1$ or $m\angle 2 + m\angle 3 = 90°$

**Exercises 1ᴓ**    Match each term to its figure.

A.  Equilateral triangle   B.  Scalene triangle      C.  Obtuse triangle
D.  Right triangle          E.  Isosceles triangle     F.  Acute triangle

**1)**

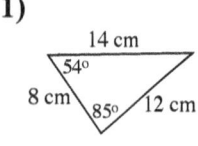

**2)**

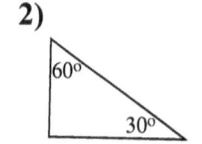

**3)**

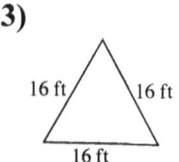

**4)**

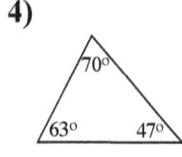

**5)**

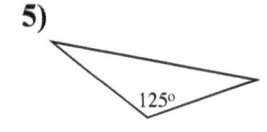

**6)**

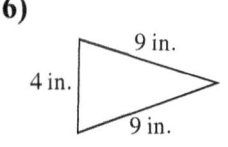

SELF-TEST

1.  Which of the following is a scalene triangle?

    **A.**  6 in., 6 in., 8 in.        **B.**  2 cm, 3 cm, 5 cm
    **C.**  4 ft, 4 ft, 4 ft          **D.**  60°, 60°, 60°

2.  Which of the following is an isosceles triangle?

    **A.**  9 cm, 6 cm, 9 cm          **B.**  7 yd, 4 yd, 9 yd
    **C.**  24 in., 24 in., 24 in.    **D.**  60°, 60°, 60°

3.  Two sides of a triangle equal 6 cm. Which of the following is the correct
    classification for the triangle?

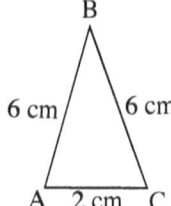

    **A.**  Scalene triangle
    **B.**  Isosceles triangle
    **C.**  Equilateral triangle
    **D.**  Right triangle

4.  A triangle has three different angles. Which of the following is the correct
    classification for the triangle?

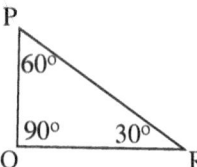

    **A.**  Scalene triangle
    **B.**  Isosceles triangle
    **C.**  Equilateral triangle
    **D.**  Right triangle

5.  A triangle has side lengths of 12 ft, 10 ft, and 8 ft. Which of the following is the
    correct classification for the triangle?

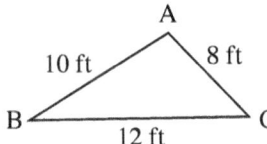

    **A.**  Scalene triangle
    **B.**  Isosceles triangle
    **C.**  Equilateral triangle
    **D.**  Right triangle

6.  All the angles of a triangle are congruent. Which of the following is the correct
    classification for the triangle?

A. Scalene triangle                              B. Isosceles triangle
C. Equiangular triangle                      D. Right triangle

7. Which of the following is the correct definition of an acute triangle?

   A. Two sides of the triangle are congruent.
   B. Two angles of the triangle are congruent.
   C. The triangle has three different sides.
   D. All three angles are less then 90°.

8. Which of the following classifies a triangle with three congruent sides?

   A. Acute triangle                           B. Scalene triangle
   C. Isosceles triangle                       D. Equilateral triangle

9. Which of the following classifies a triangle with three different sides?

   A. Acute triangle                           B. Scalene triangle
   C. Isosceles triangle                       D. Equilateral triangle

10. Which of the following classifies a triangle that has an angle that is greater than 90° and less than 180°?

    A. Acute triangle                          B. Scalene triangle
    C. Obtuse triangle                         D. Equilateral triangle

11. Which of the following classifies a triangle with at least 2 congruent sides?

    A. Acute triangle                          B. Isosceles triangle
    C. Equiangular triangle                    D. Equilateral triangle

12. Which of the following classifies the triangle shown below?

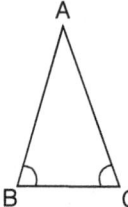

    A. Isosceles triangle
    B. Scalene triangle
    C. Equiangular triangle
    D. Equilateral triangle

## 6. Classifying Quadrilaterals

**5–12.** Classifying quadrilaterals.

> i) The sum of the measures of the interior angles of a quadrilateral is 360°.
> ii) A quadrilateral has 4 sides and 4 angles.
> iii) There are different kinds of quadrilaterals, which include, but are not limited to parallelograms, squares, rhombuses, rectangles, and trapezoids.

Properties of Parallelograms

a)  Both pairs of opposite sides are parallel.
b)  Opposite sides and angles are congruent.
c)  Diagonals bisect each other.

Properties of Isosceles Trapezoids

a)  The non-parallel sides are congruent.
b)  All base angles are congruent.
c)  The diagonals are congruent.

Properties of Rhombuses

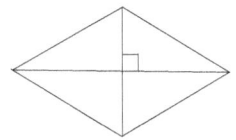

a)  All sides are congruent.
b)  The diagonals bisect the angles.
c)  The diagonals are perpendicular bisectors of each other.

Properties of Rectangles

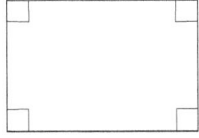

a)  Both pairs of opposite sides are parallel.
b)  All angles are congruent.
c)  Diagonals bisect each other.

Properties of Squares

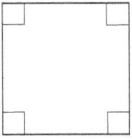

a)  All sides are congruent.
b)  All angles are congruent.
c)  Diagonals bisect each other.
d)  The diagonals are perpendicular bisectors of cach other.

**Exercises 16**    Match each term to its figure.

A. trapezoid        B. rhombus        C. rectangle        D. parallelogram

1)                    2)                    3)                    4)

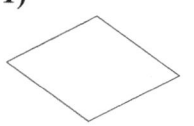     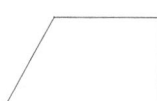     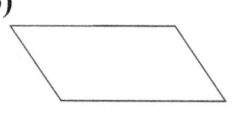

**SELF-TEST**

1. Which of the following statements is NOT a property of a rectangle?

   **A.** Diagonals that bisect each other always form four isosceles triangles.
   **B.** A rectangle has two pairs of opposite sides.
   **C.** The sum of the angles formed by the diagonals is 360°.
   **D.** A rectangle is always a parallelogram.

2. Which of the following statements can prove that ABCD is a parallelogram?

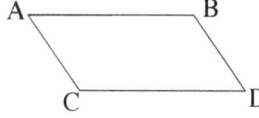

   **A.** AB and CD have parallel.
   **B.** AC and BD are not parallel.
   **C.** ∠A and ∠C are congruent.
   **D.** ∠B and ∠C are not congruent.

3. Which of the following is always a true statement?

   **A.** All of the sides of a rectangle are congruent.
   **B.** A square is a parallelogram.
   **C.** All of the angles of a rhombus are congruent.
   **D.** All of the angles of an isosceles trapezoid are congruent.

4. Which of the following is always a true statement?

   **A.** All of the angles of a rectangle are congruent.
   **B.** The sum of the inside angles of a rectangle is 360°.
   **C.** Opposite angles of a rectangle are congruent.
   **D.** All of the above.

**5.** In the figure below, ABCD is a rhombus. Which of the following is a true statement?

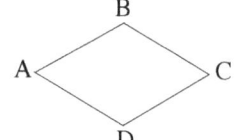

  **A.** A rhombus is a rectangle.
  **B.** A rhombus is a square.
  **C.** A rhombus is a parallelogram.
  **D.** All of the above.

**6.** Which of the following is not a true statement?

  **A.** The diagonals of a rhombus are always congruent.
  **B.** All angles of a rhombus are congruent.
  **C.** A rhombus is always a parallelogram.
  **D.** A rhombus is always a quadrilateral.

**7.** Which of the following is always a true statement?

  **A.** All sides of a rectangle are congruent.
  **B.** A square is a parallelogram.
  **C.** All angles of a rhombus are congruent.
  **D.** All of the above.

**8.** Which of the following is a true statement?

  **A.** A square is always a parallelogram.
  **B.** A square is always a quadrilateral.
  **C.** All angles of a square are always congruent.
  **D.** All of the above.

**9.** Which of the following quadrilaterals do not have perpendicular diagonals?

  **A.** Rectangle                          **B.** Square
  **C.** Rhombus                            **D.** B and C.

**10.** Which of the following quadrilaterals do not have diagonals that bisect each other?

  **A.** Rectangle                          **B.** Square
  **C.** Rhombus                            **D.** Trapezoid

**11.** Which of the following quadrilaterals do not have two pairs of opposite angles?

  **A.** Trapezoid                          **B.** Rhombus
  **C.** Rectangle                          **D.** Square

12. Which of the following quadrilaterals have four congruent sides?

    **A.** Circle                    **B.** Cylinder
    **C.** Rhombus               **D.** Isosceles trapezoid

13. Which of the following quadrilaterals always have one pair of opposite parallel sides?

    **A.** Rectangle             **B.** Parallelogram
    **C.** Rhombus               **D.** All of the above

14. Which of the following quadrilaterals always have two pairs of congruent opposite angles?

    **A.** Rectangle             **B.** Parallelogram
    **C.** Rhombus               **D.** All of the above

15. Which of the following quadrilaterals always have congruent diagonals?

    **A.** Rectangle             **B.** Square
    **C.** Rhombus               **D.** Parallelogram

16. Which of the following is a true statement about the areas of triangles and quadrilaterals?

    **A.** The area of a parallelogram is the product of its base length and its corresponding height.
    **B.** The area of a triangle is one half the product of its base length and its corresponding height.
    **C.** The area of a trapezoid is one half the product of the height and the sum of the bases.
    **D.** All of the above.

17. In the figure of a parallelogram below, which of the following is the value of $y$?

    $92°$     $88°$            **A.** $62°$          **B.** $64°$
    $114°$               **C.** $66°$          **D.** $68°$
          $y°$

# CHAPTER 6
# Probability, Statistics, and Data Analysis

> In this chapter, you will learn how to find the range, mean, median, and mode of data sets. Also, you will learn about statistics, data analysis, and probability.

## 1. Frequency and Frequency Tables

**6-1.** Know the concepts of frequency and frequency table.

> - Frequency and Frequency Tables
> - Frequency: The frequency is the number of times a data value recurs.
> - Frequency Table: A frequency table is a table that shows the total for each category or group of data.

**6-2.** 16 students in Mr. Greenfield's class took two tests during the semester. Mr. Greenfield recorded the sum of the test scores that the class had.

292, 315, 286, 252, 242, 292, 292, 292,
386, 315, 292, 286, 286, 242, 292, 292

Make a frequency table with the tally and frequency of the scores.

**SOLUTION**

a) Make a table with three columns that represents the number of test scores, the tally, and frequency.
b) First, tally the number of times a data value occurs.
c) Count the tally marks to record the frequency of each data value.

| Number of Test Scores | Tally | Frequency |
|---|---|---|
|  |  |  |
|  |  |  |
|  |  |  |
|  |  |  |
|  |  |  |
|  |  |  |

$\longrightarrow$

| Number of Test Scores | Tally | Frequency |
|---|---|---|
| 242 | \|\| | 2 |
| 252 | \| | 1 |
| 286 | \|\|\| | 3 |
| 292 | ⽧ \|\| | 7 |
| 315 | \|\| | 2 |
| 386 | \| | 1 |

**Exercises 1**    Use the information for Exercises **1-5**.
A store sells electronic items. The manager announced the results of the items sold at the end of the week.

1)    The frequency section is missing. Complete the table.

| Items | Tally | Frequency |
|---|---|---|
| Desktop Computers | ||| | |
| Laptop Computers | ~~||||~~ ~~||||~~ |||| | |
| Cell Phones | ~~||||~~ ~~||||~~ ~~||||~~ ~~||||~~ | |
| Tablets | ~~||||~~ ~~||||~~ || | |
| Televisions | ~~||||~~ ~~||||~~ ||| | |

2)    What is the greatest number of items sold in a week?

3)    What is the least number of items sold in a week?

4)    What is the difference between the number of laptop and desktop computers sold?

5)    How many tablets were sold?

**Exercises 2**    Use the information for Exercises **1-3**. The table shows the results that of the lunch items sold at school.

| Menu Sections | Tally | Frequency |
|---|---|---|
| Appetizers | ~~||||~~ |||| | 9 |
| Sandwiches | ~~||||~~ || | 7 |
| Fruits and vegetables | ||| | 3 |
| Soup | | | 1 |
| Salads | ||| | 3 |

1)    How many soup and salad items were sold?

2)    Which item was sold the least?

3)    Which item was sold the most?

## 2. Mean, Median, and Mode

**6-3.** Know the concept of mean, median, and mode, compute and compare examples to show how they may differ.

> **SOLUTION**
>
> Mean: The average number in a data set that is found by adding up the scores in the data and dividing it by the total number of scores.
> Median: The number located in the middle of a data set put in numerical order.
> Mode: The number that occurs most often in the data set.
> Range: The range is the difference between the largest and the smallest values in a data set.

**6-4.** Find the mean, median, mode, and range of the following data set;
16, 19, 17, 16, 20, 23, and 14

> **SOLUTION**
>
> a) Mean or average
>    i) Add the values. $16 + 19 + 17 + 16 + 20 + 23 + 14 = 125$
>    ii) Count how many numbers are in the data set.
>
>    16, 19, 17, 16, 20, 23, 14
>
>    **7 total in the set of this data.**
>
>    iii) Divide the sum by the total number of values in the data set.
>    $125 \div 7 = 17.86$ (Round to the nearest hundredths)
>    So, the mean of the data set is 17.86.
> b) Median or middle number
>    i) Reorder the numbers from least to greatest.
>    14, 16, 16, 17, 19, 20, 23
>    ii) Find the number in the middle.
>
>    1   2   3   4   3   2   1
>    14, 16, 16, 17, 19, 20, 23
>
>    Median or middle number
>
>    So, the median of the data set is 17.
> c) Mode
>    i) Find the number that occurs most often.
>
>    14, 16, 16, 17, 19, 20, 23
>
>    Mode
>
>    So, the mode of the data set is 16.
>    * Some data sets may not have a mode.

d) Range
   i)  Find the largest and the smallest values of the data set
       The largest value = 23 and the smallest value = 14
   ii) The difference of 23 and 14 is 9.
       So, the range in the given data set is 9.

**Exercises 3**   Find the mode and median of each data set.

1)  3, 9, 6, 4, 8, 7, 9, 8, 9, 5, 9          2)  70, 80, 95, 70, 75, 90, 70, 85, 70

3)  20, 40, 30, 40, 60, 50, 40, 40          4)  152, 164, 138, 147, 159

5)  15, 8, 10, 14, 7, 9, 13, 16             6)  24, 18, 20, 23, 20, 21, 20, 22, 20

**Exercises 4**   Find the mean and the median of each data set.

1)  126, 163, 145, 198, 178, 153, 173       2)  1, 4, 2, 7, 3, 6, 4, 2, 4

3)  25, 32, 53, 29, 35, 26, 37, 19          4)  87, 92, 84, 95, 90

5)  6, 7, 6, 9, 5, 9                        6)  53, 61, 49, 57

**Exercises 4**   Find the unknown number in given information.

1)  If the mean in the data set is 54.8, what is the unknown number?
              47, 61, _____, 68, 56

2)  If the median in the data set is 6.5, what is the unknown number?
              4, 7, _____, 9, 6, 8

3)  If the mean in the data set is 105, what is the unknown number?
              _____, 106, 98

**SELF-TEST**

1. What is the median for the data set below?
   23, 25, 27, 28, 27, 26, 26, 24

   A. 23                                    B. 24
   C. 25                                    D. 26

* Use the information below for Exercises **2-3**.
  The temperatures of the southern California coast for the next 7 days are shown in the
  following.
  82, 82, 83, 83, 85, 86, 83

2. What is the median of the temperatures?

   A. 83                                    B. 84
   C. 85                                    D. 86

3. What is the range of the temperatures?

   A. 3                                     B. 4
   C. 5                                     D. 6

4. What is the mode of the following numbers?
   190, 210, 200, 190, 190, 220, 180

   A. 190                                   B. 210
   C. 200                                   D. 180

5. Emily took a vocabulary quiz for every Friday. She noted her scores were 15, 14, 15,
   13, 15, 12, 14, 15, and 14. What is the mean of her scores?

   A. 14.1                                  B. 12.7
   C. 11.5                                  D. 15.9

6. What is the range of her scores?

   A. 2                                     B. 3
   C. 5                                     D. 6

* Use the information below for Exercises **7-9**.
  Mr. Donovan posted the scores for the math test his class took below.
  75, 77, 77, 78, 80, 81, 82, 83, 85, 86, 86, 87, 93,95, 95

7. What is the mean of the test scores?

   A. 77                                 B. 83
   C. 85                                 D. 84

8. What is the median of the test scores?

   A. 77                                 B. 83
   C. 85                                 D. 84

9. What is the range of the test scores?

   A. 25                                 B. 23
   C. 20                                 D. 19

10. The mean in the data set with the unknown number is 7. What is the unknown number?

              4, _____, 9

    A. 5                                  B. 6
    C. 7                                  D. 8

11. If the mean in the data set is increased, what is the unknown number?

              4, _____, 7

    A. 1                                  B. 3
    C. 5                                  D. 7

12. If the median in the data set is 5, what is the unknown number?

           3, 7, _____, 2, 5

    A. 1                                  B. 2
    C. 4                                  D. 8

### 3. Single-Variable Data in the Appropriate Graphs and Representations

**6–5.** Use the information below.

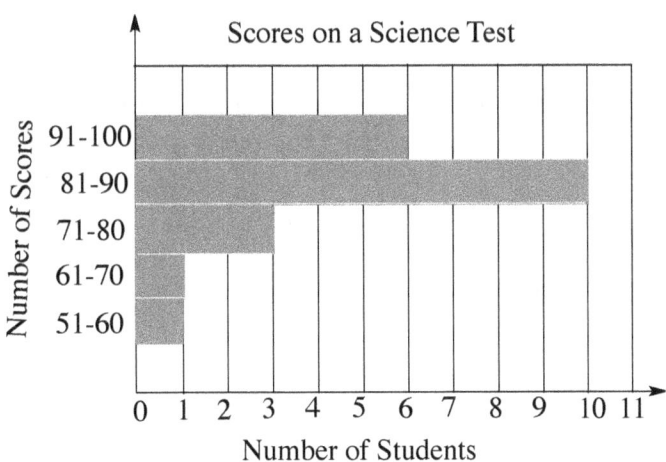

1) Find the scale of the graph
2) Find the intervals of the graph.
3) Which of the scores did only one student get?
4) Find the greatest number of scores.
5) Find the least number of scores.

> **SOLUTION**
>
> 1) The scale of a graph is the range of the values (numbers) that are 0 – 11.
> 2) The interval of a graph is the distance between two numbers on the scale. So, the interval of the graph above is 1.
> 3) The scores that only one student had were 51 – 60 and 61 – 70.
> 4) The greatest number of scores that people had is 81 – 90.
> 5) The least number of scores that people had 51 – 60 and 61 – 70.

**6–6.** Use the line plot below showing the survey of the number of books read by the class.

Weekly Reading Books

```
                    ×
                ×   ×
        ×       ×   ×   ×
    ×   ×   ×   ×   ×   ×       ×
    ×   ×   ×   ×   ×   ×       ×   ×
    _____
    1   2   3   4   5   6   7   8   9   10
```

Number of books

1) What is a line plot?
2) What is the range of the graph?
3) How many students participated in the survey?
4) How many students read more books then anyone else in the class?

**SOLUTION**

a) The line plot is a graph that shows the frequency of data along a number line. It is best to use a line plot when comparing less than 25 numbers.
b) The range of the data is the difference between the greatest and the least values in a set. So the range of this graph that is $10 - 1 = 9$
c) The total number of participating students is 22.
d) One student read 9 books.

**Exercises 6**   Use the information below for Exercises **1-5**.

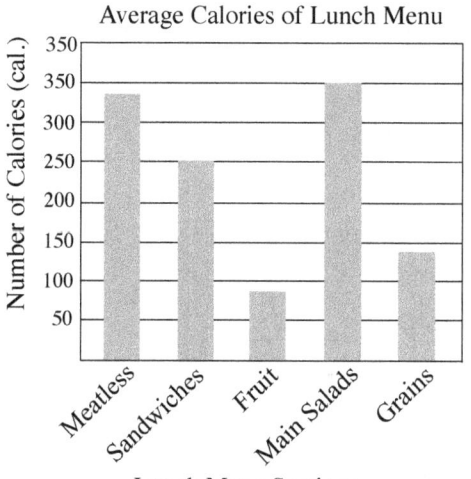

Average Calories of Lunch Menu

1) Find the scale of the graph.
2) Find the interval of the graph.
3) Find the item that has exactly 350 calories.
4) Find the item with the greatest number of calories.
5) Find the item with the least number of calories.

**Exercises 7**   Use the information shown below for Exercises **1-5**.

87, 69, 82, 76, 93, 95, 88, 99, 54, 89,
84, 92, 84, 95, 74, 83, 81, 91, 85, 88,
75

Scores on a Science Test

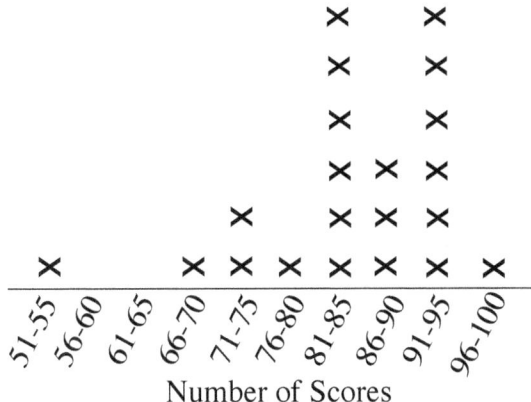

Number of Scores

1) What are the modes of the line plot?
2) What is the range of the line plot?
3) How many students have 71-75 scores?
4) What is the scale of this plot?
5) How many students took the science test?

SELF-TEST

* Use the graph for Exercises **1-3**. These information and the graph are shown the elevation of the mountains.

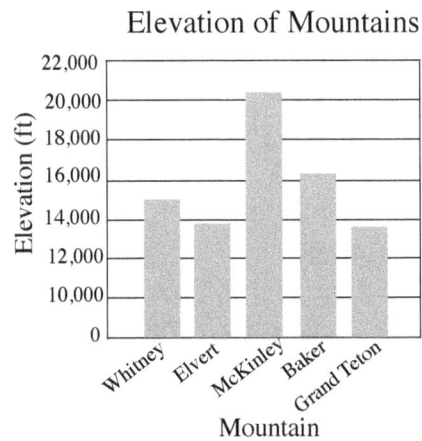

Elevation of Mountains

Mountain

1. Which mountain has the highest peak?

   **A.** Whitney        **B.** Elvert
   **C.** McKinley       **D.** Baker

2. How many mountains have a height of 16,000 ft or less?

   **A.** 1              **B.** 2
   **C.** 3              **D.** 19

**3.** Which of the following mountains have a height in the interval 12,000 ft to 14,000 ft?

  **A.** Whitney and Elvert      **B.** Whitney, Elvert, and Grand Teton
  **C.** Whitney          **D.** Elvert and Grand Teton

\* Use the graph for Exercises **4-9**. These information and the graph are shown an average rainfall in the big cities on July.

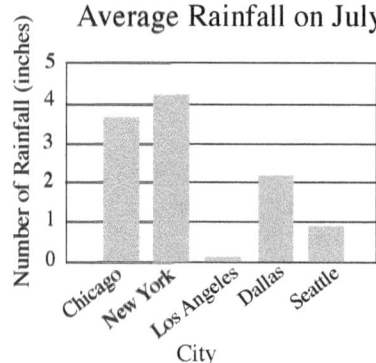

**4.** What kind of graph is shown above?

  **A.** Single-bar graph
  **B.** Line plot
  **C.** Line graph
  **D.** Double-bar graph

**5.** What is the scale of the graph?

  **A.** 0-1          **B.** 0-5
  **C.** 0-4.2         **D.** 1

**6.** What is the interval of the graph?

  **A.** 1          **B.** 2
  **C.** 3          **D.** 4

**7.** What is the range of the average rainfall for these five cities?

  **A.** 1          **B.** 2
  **C.** 3          **D.** 4

**8.** Which of the cities had 4 inches or more rainfall in July?

  **A.** Chicago        **B.** New York
  **C.** Los Angeles      **D.** Seattle

**9.** Which city had the highest rainfall in July?

      **A.** Chicago                  **B.** New York
      **C.** Dallas                    **D.** Seattle

\* Use the line plot below.
Number of hits made by the home team

```
                ×
                ×
     ×          ×
     ×   ×    × × ×       ×
    _____
     1  2  3  4  5  6  7  8  9
          Number of innings
```

**10.** What was the number of hits made during the 5ᵗʰ inning?

      **A.** 1                 **B.** 2
      **C.** 3                 **D.** 4

**11.** Which inning had two points?

      **A.** 2                 **B.** 3
      **C.** 5                 **D.** 6

**12.** What is the total number of hits made for the home team?

      **A.** 1                    **B.** 2
      **C.** 4                    **D.** 10

**13.** What is the range of the baseball scores?

      **A.** 4                    **B.** 5
      **C.** 7                    **D.** 8

## 4.  Know how to correctly write ordered pairs; for example (x, y)

**6–7.** The bowling scoring system was showed the scores on the frame below.

| Frame | 1 | 2 | 3 | 4 | 5 | 6 | 7 | 8 | 9 | 10 |
|-------|---|---|---|---|---|----|---|---|---|----|
| Score | 4 | 7 | 6 | 9 | 4 | 13 | 5 | 8 | 7 | 6 |

    **1)**  Find the range of the scores.
    **2)**  Find the median of the scores.
    **3)**  Make an appropriate graph of the ordered pairs.

**SOLUTION**

1) The range is the difference between the largest and the smallest values in a set of data. So, the range of the score is 13 − 4 = 9.
2) The median of the score is 6.5
3) Make the graph for the ordered pairs.

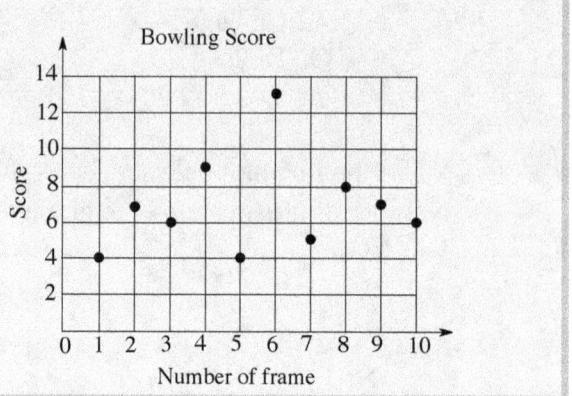

**Exercises 8**    Use the table and the graph for Exercises **1-7**. The table below shows the number of recorded storms in Florida, 2013.

| Months | Apr | May | June | July | Aug | Sep | Oct | Nov | Dec |
|---|---|---|---|---|---|---|---|---|---|
| Number of Storms | 1 | 2 | 7 | 8 | R | 17 | 8 | 2 | S |

The ordered pairs show the number of recorded storms in Florida.

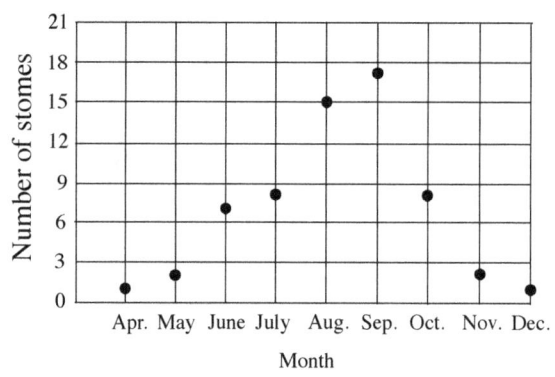

Number of Recorded Stoms in Florida, 2013

1) Find the scales of the graph

2) Determine the values of R and S in the table.

3) Find the median of the number of the storms.

4) What is the range of the graph?

5) Find the mean of the number of the storms.

6) How many storms occurred between September and November?

7) How many storms occurred between May and September?

## 5.  Various Graphs

\*  Double-bar graphs are used to compare more than one kind of data.

**Exercises 9**    The table below shows a set of data about recyclables collected every
week.

Weekly Collected Recyclables

| Week | 1 week | 2 week | 3 week | 4 week |
|---|---|---|---|---|
| Plastic Bottles | 55 | 29 | 31 | 45 |
| Cans | 42 | 32 | 24 | 36 |

\*  Make a double-bar graph that uses that data set of the table above. Label everything

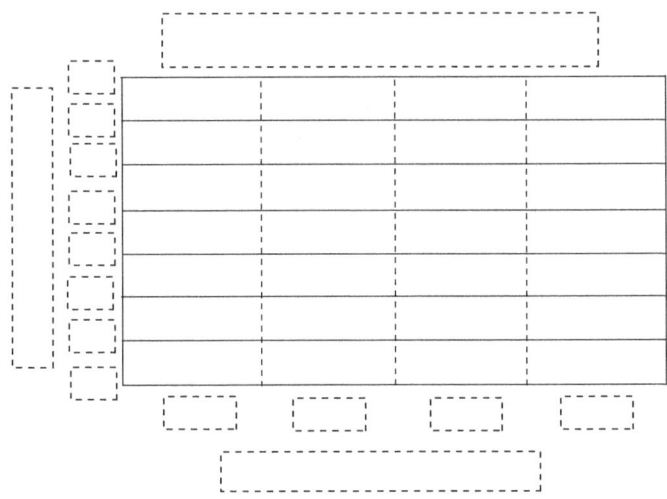

1)  Which week had the greatest number of recyclables collected?

2)  Which week had the least number of recyclables collected?

3)  Which week had the least number of cans collected?

4)  Describe your thoughts based on the double-bar graph.

**Exercises 10**    For Exercises **1-6**, the information below shows the average rainfall in Memphis and Minneapolis during January, April, and August.

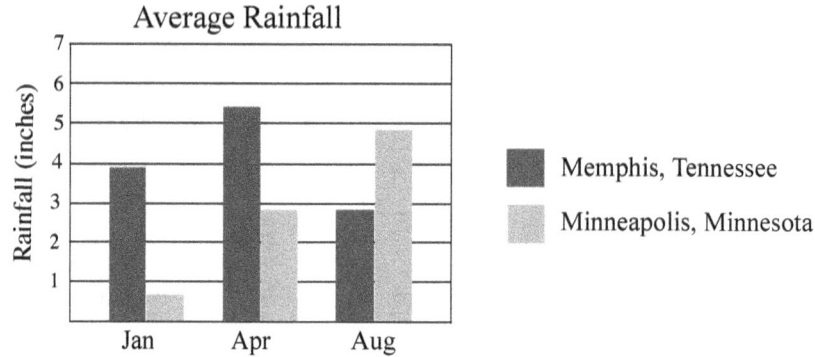

1)  Which city had the highest rainfall?

2)  Which month had the greatest number of rainfall in total?

3)  Which month had the lowest rainfall for Memphis?

4)  Which month had the lowest rainfall for Minneapolis?

5)  Which month had the greatest difference of the average rainfall between the two cities?

6)  Describe your thoughts based on the double-bar graph.

**Exercises 11**    The graph below shows the monthly electric bills for a year.

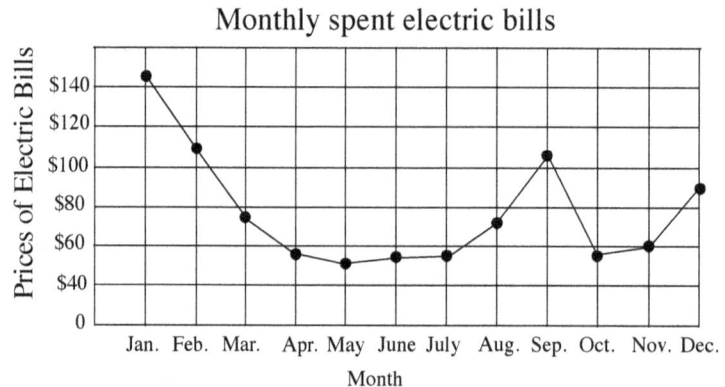

**1)**  What kind of graph is shown above?

**2)**  Which month had the highest electric bill?

**3)**  Which month had the lowest bill?

**4)**  Which months had bills that exceeded $100?

**5)**  What is the difference between the bills paid in July and September?

**6)**  What is the difference between the bills paid in January and December?

**7)**  Describe your thoughts based on the line graph

**Exercises 12**    The graph below shows the average temperature in Iowa City, Iowa.

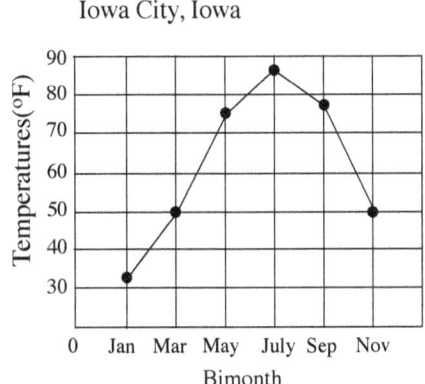

Bimonthly average temperature in
Iowa City, Iowa

**1)**  Which month had the greatest temperature?

**2)**  What is the scale of the line graph?

**3)**  What is the interval of the graph?

**4)**  During which month did the temperature start to drop?

**5)**  During which months did the temperature average around 50 °F?

**6)**  Give an estimate of what the temperature will be in December based on the line graph.

* For Exercises **1-5**, the information and the graph show the time spent playing a video game.

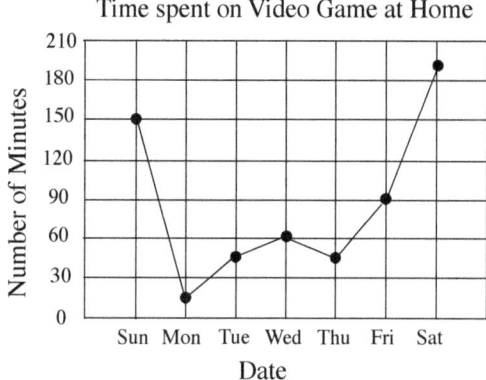

1. What kind of graph is shown above?

   **A.** Single-bar graph          **B.** Line plot
   **C.** Line graph               **D.** Double-bar graph

2. What day had the greatest amount of time spent playing?

   **A.** Sun.                     **B.** Mon
   **C.** Wed.                     **D.** Sat.

3. What is the scale of the line graph?

   **A.** 0-210                    **B.** 0-30
   **C.** 30                       **D.** 15-190

4. What is the interval of the graph?

   **A.** 0-210                    **B.** 0-30
   **C.** 30                       **D.** 15-190

5. When did a sharp increase in the playing time occur?

   **A.** Sun-Mon                  **B.** Fri-Sat
   **C.** Mon-Tue                  **D.** Sat-Sun

* Use the information below for Exercises **6-9**.
  Eric would like to check his weight for the next 12 months.

| Number of Months | 0 | 2 | 4 | 6 | 8 | 10 | 12 |
|---|---|---|---|---|---|---|---|
| Number of Weights (Ib) | 73 | 73 | 74 | 75 | 76 | 76 | 77 |

6. What kind of graph is appropriate for the data set above?

   A. Single-bar graph              B. Line plot
   C. Line graph                    D. Double-bar graph

7. What is the range of the data set?

   A. 1                             B. 2
   C. 3                             D. 4

8. Which of the following numbers is the median of the data?

   A. 73                            B. 74
   C. 75                            D. 76

9. What is the difference between the data recorded in the $4^{th}$ and $10^{th}$ months?

   A. 1                             B. 2
   C. 3                             D. 4

\* Use the table below for Exercises **10-14**. Anderson recorded the hours he spent working on his math and English.

| Week | 1 | 2 | 3 | 4 |
|---|---|---|---|---|
| English | 5 | 6 | 5 | 3 |
| Math | 7 | 9 | 5 | 4 |

10. What kind of graph would be best for this data set?

    A. Single-bar graph             B. Line plot
    C. Line graph                   D. Double-bar graph

11. Which week did he spent the least amount of time on his English homework?

    A. 1                            B. 2
    C. 3                            D. 4

**12.** Which week did he spent the greatest amount of time on his homework?

    **A.** 1                     **B.** 2
    **C.** 3                     **D.** 4

**13.** What is the range of the time he spent on his math homework?

    **A.** 2                     **B.** 3
    **C.** 4                     **D.** 5

**14.** A report displays your growing progress, from the time of your birth to the present. What graph would be ideal for this?

    **A.** Single-bar graph           **B.** Line plot
    **C.** Line graph                 **D.** Double-bar graph

# ANSWERS

## CHAPTER 1

### Exercises 1
1) Two hundred fifty-three
   200 + 50 + 3
2) Eight-three thousand, seven hundred forty-one
   80,000 + 3,000 + 700 + 40 + 1
3) Sixty-nine thousand, four
   60,000 + 9,000 + 4
4) Four million, two hundred seventy thousand, two hundred seventy
   4,000,000 + 200,000 + 70,000 + 200 + 70
5) Five hundred thousand, thirty-six
   500,000 + 30 + 6

### Exercises 2
| | | | | | | | |
|---|---|---|---|---|---|---|---|
| 1) | 3,286 | 2) | 685,031 | 3) | 295,000 | 4) | 535,216,015 |

### Exercises 3
| | | | | | | | |
|---|---|---|---|---|---|---|---|
| 1) | 71,520 | 2) | 1,200 | 3) | 295.500 | 4) | 1,700 |
| 5) | 84,000,000 | 6) | 76,080 | 7) | 64,000 | 8) | 721.01 |
| 9) | 50,000,000 | 10) | 70,000,000 | 11) | 36,400,000 | 12) | 50,300,000 |

### Exercises 4
| | | | | | | | |
|---|---|---|---|---|---|---|---|
| 1) | 35,000 | 2) | 380,000 | 3) | 250,000 | 4) | 736,000 |
| 5) | 700,000 | 6) | 42,700,000 | 7) | 36,400,000 | 8) | 263,000,000 |
| 9) | 500,000 | 10) | 50,000 | 11) | 465,000 | 12) | 110,000 |

### Exercises 5
| | | | | | | | |
|---|---|---|---|---|---|---|---|
| 1) | 15,000 | 2) | 16,000 | 3) | Thousands, B | 4) | 800 |
| 5) | J | 6) | hundreds | | | | |

### Exercises 6
| | | | | | | | |
|---|---|---|---|---|---|---|---|
| 1) | 118,000 | 2) | 40,000 | 3) | 30,000 | 4) | 20,000 |
| 5) | 670,000 | 6) | 20,000 | 7) | 370,000 | 8) | 130,000 |
| 9) | 0 | 10) | 970,000 | 11) | 50,000 | 12) | 40,000 |

### Exercises 7
| | | | | | | | |
|---|---|---|---|---|---|---|---|
| 1) | 386,257 | 2) | 84,498 | 3) | 22,588 | 4) | 847,005 |
| 5) | 464,941 | 6) | 42,559 | 7) | 44,057 | 8) | 199,438 |
| 9) | 152,743 | | | | | | |

### Exercises 8
| | | | | | | | |
|---|---|---|---|---|---|---|---|
| 1) | 2,762 | 2) | 539 | 3) | 3,061 | 4) | 7,447 |
| 5) | 5,552 | 6) | 14,372 | 7) | 9,472 | 8) | 11,880 |
| 9) | 799 | 10) | 4,343 | | | | |

### Exercises 9
| | | | | | | | |
|---|---|---|---|---|---|---|---|
| 1) | 1,000 | 2) | 40 | 3) | 9,000 | 4) | 30,000 |
| 5) | 40 or 50 | 6) | 90 or 100 | 7) | 4,000 or 5,000 | 8) | 60 or 70 |

**Exercises 10**
1) 900,000　　2) 20 or 30　　3) 3,200,000　　4) 40 or 50
5) 20 or 30　　6) 60,000　　7) 2,100　　8) 9,000
9) 180　　10) 70 or 80　　11) 18,000　　12) 560,000

**Exercises 11**
1) 3,000　　2) 9,000　　3) 1,300

**Exercises 12**
1) 625,091　2) 726,091　3) 674,908　4) 1,158,199　5) 733,584　6) 1,032,833
7) 168,496　8) 346,323　9) 1,366,555

**Exercises 13**
1) 16,603　2) 6,961　3) 15,652　4) 24,632　5) 11,497　6) 14,347
7) 6,949　8) 4,472

**Exercises 14**
1) 266,889　2) 388,356　3) 268,269　4) 188,947　5) 66,693　6) 135,643
7) 67,579　8) 173,724　9) 477,789　10) 677,898　11) 350,060　12) 269,963

**Exercises 15**
1) 19,833　2) 21,249　3) 89,026　4) 17,537　5) 19,116　6) 100,268
7) 45,548　8) 64,211

**Exercises 16**
1) 538,335　2) 360,352　3) 591,147　4) 420,035　5) 184,710　6) 422,388
7) 293,432　8) 723,200　9) 560,064　10) 260,778

**Exercises 17**
1) 9　2) 8　3) 53　4) 68　5) 8　6) 720
7) 6　8) 9

**Exercises 18**
1) 5,681　2) 12,735　3) 50,589　4) 21,175　5) 16,165　6) 37,884
7) 6,552　8) 18,080　9) 6,888　10) 35,208　11) 49,950

**Exercises 19**
1) 32　2) 45　3) 86　4) 48　5) 12　6) 30
7) 28　8) 34

**Exercises 20**
1) 83　2) 45　3) 34　4) 8　5) 13　6) 31
7) 68　8) 104　9) 76　10) 64　11) 59

**Exercises 21**
1) 8　2) 6　3) 800　4) 559　5) 9　6) 378
7) 16　8) 3,348

**Exercises 22**
1) 11,587　2) 643r4　3) 1,463r21　4) 763r9　5) 1,895r13　6) 850r34
7) 665r12　8) 2,512r37　9) 96r16　10) 57r8　11) 49r8

### Exercises 23
1) 64    2) 94    3) 3,952    4) 1,805    5) 38    6) 3,936
7) 57    8) 3,572

### Exercises 24
1) 9,692    2) 3,009    3) 2,411    4) 6,465    5) 10,784    6) 11,841

### Exercises 25
1) 152    2) 81    3) 21    4) In 2012 = 106,096 and in 2013 = 13,262
5) 496, 248    6) $1031.00, $1043.00    7) $19.00

### Exercises 26
1) Multiply the difference of three and four with the sum of three and four
2) The sum of four and the quotient of three and two subtracted by one
3) The quotient of eight and four minus three
4) The product of six and seven increased by eight
5) Multiply the difference of seven and three by four minus two

### Exercises 27
1) $(5-3) \times 6$    2) $(9+2) \times 2 + 4$    3) $10 \div 2 - 2$    4) $(8+2) + (9 \div 3)$
5) $(2+8) - (8 \times 2)$

### Exercises 28
1) 105    2) 770    3) 35    4) 26    5) 26    6) 7
7) 16    8) 53    9) 24    10) 6

### Exercises 29
1) 62    2) 10    3) 6.99    4) 18    5) 13    6) 16
7) 18    8) 12

### Exercises 30
1) 18, 10    2) 866

## CHAPTER 2

### Exercises 1
1) 21    2) 28    3) 2    4) 1    5) 0    6) 2
7) 1    8) 2    9) 6    10) 1

### Exercises 2
1) 25    2) 15    3) 24    4) 10    5) 19    6) 2
7) 23    8) 23    9) 8    10) 10

### Exercises 3
1) 0.7    2) 7.1    3) 48.6    4) 5.25    5) 8.8    6) 9.55
7) 6.05    8) 2.2    9) 5.6    10) 3.2

### Exercises 4
1) 10    2) 10    3) 5    4) 10    5) 3    6) 2.6
7) 10    8) 9    9) 2    10) 4

### Exercises 5
1) 10    2) 21    3) 28    4) 34    5) 27    6) 4

### Exercises 6
1) 28  2) 4  3) 1  4) 5  5) 21  6) 12
7) 11  8) 8  9) 11  10) 27  11) 9  12) 39
13) 37

### Exercises 7
1) -3.5  2) 4.05  3) 12.1  4) -3.4  5) 23.5  6) -3.9
7) -0.2  8) 7.6

### Exercises 8
1) 19  2) 2  3) 7  4) 6  5) 4  6) 24
7) 20  8) 8  9) 12  10) 8.8

### Exercises 9
1) 10  2) 36  3) 10  4) 17  5) 4  6) 18
7) 9  8) 5  9) 9  10) 11.5  11) 1  12) 6.6

### Exercises 10
1) 4  2) 55  3) 12  4) 6  5) 1  6) 31
7) 67  8) 206  9) 41  10) 19

### Exercises 11
1) 3  2) 4  3) 2  4) 53  5) 8  6) 4
7) 37.2  8) 50  9) 8  10) 4

### Exercises 12
1) 12  2) 22  3) 26.875  4) 12.5  5) 7  6) 13
7) 4  8) 3  9) 4  10) 3

### Exercises 13
1) 15  2) 5  3) 2  4) 7  5) 4  6) 10
7) 18  8) 3  9) 2  10) 18

### Exercises 14
1) 9  2) 24  3) 28  4) 2  5) 15  6) 3
7) 8.3  8) 2  9) 16  10) 6

### Exercises 15
1) 6  2) 4  3) 4  4) 9  5) 4  6) 11
7) 48  8) 5.6  9) 16  10) 2

### Exercises 16
1) 1/2  2) 2  3) 18.6  4) 8  5) 35  6) 2.8
7) 5  8) 27  9) 27  10) 2

### Exercises 17
1) 3  2) 4  3) 1  4) 2  5) 4  6) 17
7) 6  8) 36  9) 7  10) 44

### Exercises 18
1) 36  2) 107  3) 180  4) 80  5) 41  6) 15
7) 8  8) 44.5  9) 10  10) 24  11) 7  12) 6

### Exercises 19
1) 13   2) 15   3) 35   4) 42   5) 20   6) 2
7) 5.25   8) 27   9) 6   10) 10

### Exercises 20
1) 25   2) 36   3) 13   4) 7   5) 2.5   6) 21
7) 6.5   8) 16

### Exercises 21
1) 39   2) 98   3) 24   4) 15   5) 6   6) 56
7) 15   8) 32   9) 15   10) 15

### Exercises 22
1) 3   2) 12   3) 10   4) 6   5) 8   6) 8
7) 6(3/4)   8) 2

### Exercises 23
1) – 4)

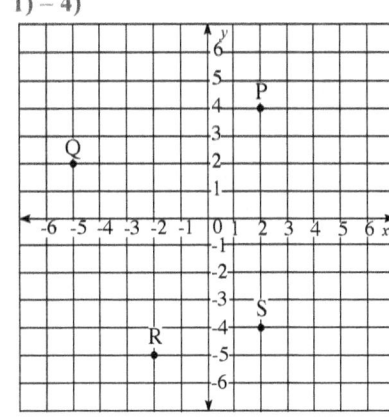

### Exercises 24
1) (-4, -3)   2) (0, -5)   3) (4, -3)   4) (4, 3)   5) (0, 5)   6) (-4, 3)

### Exercises 25
1) Q   2) (4, 2)   3) P

### Exercises 26
1) E   2) (-3, -5)   3) A

### Exercises 27
1) 4   2) 3

### Exercises 28
1) -3   2) (-1, 2) (7, -4)

### Exercises 29
1) A(4, 4), B(4, 10)   2) 6 units

### Exercises 30
1) C(3, 15), D(18, 15)   2) 15 units

## Exercises **31**

1)    B      2)    The point can be moved forward to 4 units and downward to 4 units (B → D).

## Exercises **32**

1)    6 units    2)    4 units    3)    4 units    4)    7 units

## Exercises **33**

|   | A | B | C | D |
|---|---|---|---|---|
| $x$ | 0 | 2 | 4 | 6 |
| $y$ | 4 | 6 | 8 | 10 |

## Exercises **34**

1)    7 units    2)    6 units    3)    7 units    4)    6 units    5)    26 units

## Exercises **35**

1)    3 units    2)    3 units    3)    9 units    4)    9 units    5)    Rectangle    6)    24 units

## Exercises **36**

1)

| $x$ | −1 | 0 | 1 | 2 |
|---|---|---|---|---|
| $y$ | −1 | 0 | 1 | 2 |

2)

| $x$ | 0 | 1 | 2 | 3 |
|---|---|---|---|---|
| $y$ | 0 | −2 | −4 | −6 |

3)

| $x$ | −3 | −1 | 1 | 3 |
|---|---|---|---|---|
| $y$ | 4 | 2 | 0 | −2 |

4)

| $x$ | 2 | 4 | 6 | 8 |
|---|---|---|---|---|
| $y$ | 3 | 7 | 11 | 15 |

5)

| $x$ | −1 | −2 | −3 | −4 |
|---|---|---|---|---|
| $y$ | 1 | 3 | 5 | 7 |

6)

| $x$ | −2 | −1 | 0 | 1 |
|---|---|---|---|---|
| $y$ | −3 | −1 | 1 | 3 |

## Exercises **37**

1)

| $x$ | −2 | 0 | 2 | 4 |
|---|---|---|---|---|
| $y$ | −4 | 2 | 8 | 16 |

2)

| $x$ | −1 | 0 | 1 | 2 |
|---|---|---|---|---|
| $y$ | −5 | −1 | 3 | 7 |

3)

| $x$ | −1 | 0 | 1 | 2 |
|---|---|---|---|---|
| $y$ | −1 | 1 | 3 | 5 |

4)

| $x$ | 0 | 1 | 2 | 3 |
|---|---|---|---|---|
| $y$ | −3 | −2 | −1 | 0 |

5)

| $x$ | 3 | 2 | 1 | −4 |
|---|---|---|---|---|
| $y$ | −2 | −1 | 0 | 5 |

6)

| $x$ | −9 | −3 | 3 | 27 |
|---|---|---|---|---|
| $y$ | −3 | −1 | 1 | 9 |

## Exercises **38**

1)

| $x$ | 8 | 4 | 0 | −20 |
|---|---|---|---|---|
| $y$ | −2 | −1 | 0 | 5 |

2)

| $x$ | −10 | −5 | 8 | 25 |
|---|---|---|---|---|
| $y$ | −2 | −1 | 0 | 5 |

3)

| $x$ | −6 | −2 | 2 | 18 |
|---|---|---|---|---|
| $y$ | −3 | −1 | 1 | 9 |

4)

| $x$ | −2 | 0 | 2 | 12 |
|---|---|---|---|---|
| $y$ | −2 | −1 | 0 | 5 |

5)

| $x$ | −9 | −6 | −3 | 12 |
|---|---|---|---|---|
| $y$ | −2 | −1 | 0 | 5 |

6)

| $x$ | 3 | 2 | 1 | −4 |
|---|---|---|---|---|
| $y$ | −2 | −1 | 0 | 5 |

### Exercises 39

**1)**

| $x$ | 5 | 4 | 3 | −2 |
|---|---|---|---|---|
| $y$ | −2 | −1 | 0 | 5 |

**2)**

| $x$ | −10 | −5 | 0 | 25 |
|---|---|---|---|---|
| $y$ | −2 | −1 | 0 | 5 |

**3)**

| $x$ | 1 | −1 | −3 | −13 |
|---|---|---|---|---|
| $y$ | −2 | −1 | 0 | 5 |

**4)**

| $x$ | −4 | 0 | 4 | 20 |
|---|---|---|---|---|
| $y$ | −3 | −1 | 1 | 9 |

**5)**

| $x$ | −10 | −8 | −6 | 4 |
|---|---|---|---|---|
| $y$ | −2 | −1 | 0 | 5 |

**6)**

| $x$ | −1 | 1 | 3 | 11 |
|---|---|---|---|---|
| $y$ | −3 | −1 | 1 | 9 |

## CHAPTER 3

### Exercises 1

1) Five hundredths
   Expanded Form: $\frac{5}{100}$
2) Two and six hundredths
   $2 + \frac{6}{100}$
3) Eighty-three and eight hundredths
   $80 + 3 + \frac{8}{100}$
4) Two hundred forty-three and four hundred twenty-nine thousandths
   $200 + 40 + 3 + \frac{4}{10} + \frac{2}{100} + \frac{9}{1000}$
5) Seven tenths
   Expanded Form: $\frac{7}{10}$
6) Seven hundred seven thousandths
   Expanded Form: $\frac{7}{10} + \frac{7}{1000}$

### Exercises 2

1) 609.501  **2)** 0.006  **3)** 5.26  **4)** 200.1

### Exercises 3

| 1) | $3.00 | 2) | 53 | 3) | 708 | 4) | 123 | 5) | 16 | 6) | 91 |
|---|---|---|---|---|---|---|---|---|---|---|---|
| 7) | 74 | 8) | $443 | 9) | 67 | 10) | 24 | 11) | 2 | 12) | 64 |

### Exercises 4

| 1) | < | 2) | < | 3) | < | 4) | = | 5) | = | 6) | < |
|---|---|---|---|---|---|---|---|---|---|---|---|
| 7) | < | 8) | > | 9) | < | 10) | < | 11) | < | 12) | = |

### Exercises 5

1) $2\frac{1}{8} > 2.11 > 2\frac{1}{10} > 2.05$
2) $10.4 > 0.95 > -5.09 > -10.99$
3) $2\frac{1}{3} > 2.27 > 2\frac{1}{4} > 2.17 > 2.15$
4) $\frac{8}{5} > 0.392 > 0.387 > 0.38 > 0.359$
5) $92.6489 > 92.648 > 91.998 > 91\frac{4}{5}$
6) $\frac{36}{10} > 2\frac{36}{100} > 2.35 > 2.3 > 2.05$

### Exercises 6

| 1) | 19.4 | 2) | 20.1 | 3) | 21.5 | 4) | 21.8 | 5) | 11.5 | 6) | 11.75 |
|---|---|---|---|---|---|---|---|---|---|---|---|
| 7) | 12.5 | 8) | 13.25 | | | | | | | | |

**Exercises 7**

| | | | | | | | | | | | |
|---|---|---|---|---|---|---|---|---|---|---|---|
| **1)** | 1.29 | **2)** | 5.55 | **3)** | 40.00 | **4)** | 16.1 | **5)** | $39.60 | **6)** | 2.818 |
| **7)** | 6.2 | **8)** | 15.66 | **9)** | 13.10 | | | | | | |

**Exercises 8**

| | | | | | | | |
|---|---|---|---|---|---|---|---|
| **1)** | 0.89 | **2)** | 1.94 | **3)** | 2.79 | **4)** | 1.29 |

**Exercises 9**

| | | | | | | | | | | | |
|---|---|---|---|---|---|---|---|---|---|---|---|
| **1)** | 5.3 | **2)** | 29.46 | **3)** | 6.05 | **4)** | 6.92 | **5)** | 1.02 | **6)** | 59.99 |
| **7)** | 69.32 | **8)** | 15.401 | **9)** | 12.75 | **10)** | 4.002 | **11)** | 4 | **12)** | 7 |

**Exercises 10**

| | | | | | | | | | | | |
|---|---|---|---|---|---|---|---|---|---|---|---|
| **1)** | 0.77 | **2)** | 0.32 | **3)** | 2.16 | **4)** | 0.67 | **5)** | 6.06 | **6)** | 0.59 |
| **7)** | 1.06 | **8)** | 3.16 | **9)** | 1.75 | **10)** | 0.27 | | | | |

**Exercises 11**

| | | | | | | | | | | | |
|---|---|---|---|---|---|---|---|---|---|---|---|
| **1)** | 5.2 | **2)** | 0.94 | **3)** | 2.78 | **4)** | 2.03 | **5)** | 3.61 | **6)** | 7.31 |
| **7)** | 2.4 | **8)** | 2.61 | | | | | | | | |

**Exercises 12**

| | | | | | | | | | | | |
|---|---|---|---|---|---|---|---|---|---|---|---|
| **1)** | 39.5 | **2)** | 8.3 | **3)** | $8.50 | **4)** | $0.60 | **5)** | 30.16 | **6)** | 7.15 |
| **7)** | $0.69 | **8)** | $2.82 | **9)** | $11.48 | **10)** | $11.85 | **11)** | 9.8 | **12)** | 8.45 |
| **13)** | 1.7 | | | | | | | | | | |

**Exercises 13**

| | | | | | | | | | | | |
|---|---|---|---|---|---|---|---|---|---|---|---|
| **1)** | 5.55 | **2)** | $13.25 | **3)** | 0.76 | **4)** | $38.63 | **5)** | 0.41 | **6)** | 3.19 |

**Exercises 14**

| | | | | | | | | | | | |
|---|---|---|---|---|---|---|---|---|---|---|---|
| **1)** | 0.23 | **2)** | 1.68 | **3)** | 5.26 | **4)** | 2.47 | **5)** | 1.94 | **6)** | 6.31 |

**Exercises 15**

| | | | | | | | | | | | |
|---|---|---|---|---|---|---|---|---|---|---|---|
| **1)** | 2.2 | **2)** | 2.50 | **3)** | 4.22 | **4)** | 1.85 | **5)** | 2.14 | **6)** | 1.27 |

**Exercises 16**

| | | | | | | | | | | | |
|---|---|---|---|---|---|---|---|---|---|---|---|
| **1)** | 0.4 | **2)** | 2.4 | **3)** | 32.5 | **4)** | 4.8 | **5)** | 35.2 | **6)** | 60.8 |

**Exercises 17**

| | | | | | | | | | | | |
|---|---|---|---|---|---|---|---|---|---|---|---|
| **1)** | 60.48 | **2)** | $153.75 | **3)** | 17.52 | **4)** | 44.0 | **5)** | 11.85 | **6)** | 2.36 |

**Exercises 18**

| | | | | | | | | | | | |
|---|---|---|---|---|---|---|---|---|---|---|---|
| **1)** | 3 | **2)** | 3 | **3)** | 2.2 | **4)** | 3 | **5)** | 2.5 | **6)** | 5.5 |

**Exercises 19**

| | | | | | | | | | | | |
|---|---|---|---|---|---|---|---|---|---|---|---|
| **1)** | 7.2 | **2)** | 3.075 | **3)** | 1.4 | **4)** | 24.3 | **5)** | 14.7 | **6)** | 8.84 |

**Exercises 20**

| | | | | | | | | | | | |
|---|---|---|---|---|---|---|---|---|---|---|---|
| **1)** | 6.52 | **2)** | 76.3 | **3)** | $21.25 | **4)** | $49.37 | **5)** | 162.36 | **6)** | $76.70 |
| **7)** | 76.08 | **8)** | 47.7 | **9)** | 57.96 | **10)** | 0.3 | **11)** | 3.435 | **12)** | 44.54 |

**Exercises 21**

| | | | | | | | | | | | |
|---|---|---|---|---|---|---|---|---|---|---|---|
| **1)** | 0.31 | **2)** | 9.3 | **3)** | 2.01 | **4)** | 142 | **5)** | 10.5 | **6)** | 60 |
| **7)** | 15 | **8)** | 0.41 | **9)** | 3.2 | **10)** | 50 | **11)** | 1.8 | **12)** | 6 |
| **13)** | 6.16 | **14)** | 0.47 | **15)** | 7.9 | | | | | | |

### Exercises 22
1) 0.3    2) 1.9    3) 5.6    4) 0.4    5) 0.8    6) 0.5
7) 0.81    8) 0.44

### Exercises 23
1) 4.386    2) 6.4    3) 5.09    4) 0.238    5) 0.36    6) 2.13

### Exercises 24
1) 66.8    2) 5.18    3) 6.6    4) 21    5) 2.23    6) 1.34
7) 5.71

### Exercises 25
1) 1.4    2) 0.48    3) 15.435    4) 9.016    5) 1.4    6) 13.56

### Exercises 26
1) 8.64    2) 1.5    3) 8    4) 3.62    5) 0.84    6) 1.8

### Exercises 27
1) Thirteen hundredths    2) Two and seven eighths
3) Two and four fifths    4) Eighty-nine hundredths

### Exercises 28
1) 4/9    2) 10(2/7)    3) 6/11    4) 3(5/12) – (6/8)

### Exercises 29
1) <    2) <    3) >    4) <    5) >    6) >
7) >    8) <

### Exercises 30
1) <    2) >    3) =    4) =    5) >    6) <
7) >    8) >    9) <    10) >

### Exercises 31
1) $2\frac{1}{2} > 2\frac{1}{3} > 2\frac{1}{4} > 2\frac{1}{5} > 2\frac{1}{6}$

2) $2.85 > 2\frac{5}{6} > 2\frac{4}{5} > 2\frac{3}{4} > 2\frac{2}{3}$

3) $2\frac{33}{50} > 2\frac{16}{25} > 2\frac{3}{5} > 2\frac{11}{100} > 2\frac{1}{10}$

4) $2\frac{15}{40} > 2\frac{45}{100} > 2\frac{1}{4} > 2.24 > 2\frac{4}{20}$

5) $\frac{7}{10} > \frac{10}{15} > 2\frac{9}{30} > \frac{1}{4} > \frac{1}{5} >$

6) $0.34 > \frac{3}{9} > 0.306 > \frac{2}{7} > \frac{1}{4}$

### Exercises 32
1) 5(3/4)    2) 6(1/4)    3) 7    4) 7(1/4)

### Exercises 33
1) 2(1/3)    2) 3    3) 3(1/3)    4) 4(1/3)

### Exercises 34
1) $15\frac{2}{5}$    2) $15\frac{4}{5}$    3) $16\frac{2}{5}$    4) $16\frac{3}{5}$

### Exercises 35

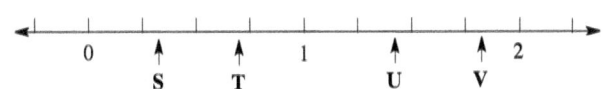

# Answers

**Exercises 36**
1) $2\frac{1}{2}$   2) $(\frac{3}{4})$   3) $(\frac{39}{100})$   4) $(\frac{1}{10})$   5) $1\frac{1}{5}$   6) $1\frac{1}{20}$
7) $2\frac{1}{20}$   8) $(\frac{1}{250})$   9) $(\frac{9}{1000})$   10) $20\frac{101}{1000}$

**Exercises 37**
1) 0.125   2) 0.25   3) 0.4   4) 0.8   5) 0.4   6) 2.2
7) 0.4375   8) 5.75

**Exercises 38**
1) $1\frac{2}{5}$   2) $4\frac{1}{4}$   3) 4   4) $6\frac{1}{7}$   5) $2\frac{3}{5}$   6) 5
7) $(\frac{7}{10})$   8) 1   9) $5\frac{1}{3}$   10) $(\frac{4}{5})$   11) $2\frac{1}{4}$   12) $2\frac{1}{8}$

**Exercises 39**
1) $5\frac{5}{6}$   2) $2\frac{2}{3}$   3) $3\frac{2}{3}$   4) 8   5) 1   6) 2

**Exercises 40**
1) $1\frac{2}{35}$   2) $2\frac{5}{9}$   3) $4\frac{1}{4}$   4) $5\frac{9}{10}$   5) $2\frac{1}{4}$   6) 4
7) $5\frac{1}{8}$   8) $3\frac{4}{15}$   9) $4\frac{1}{6}$   10) $3\frac{1}{12}$   11) $3\frac{14}{15}$   12) $2\frac{9}{28}$

**Exercises 41**
1) $3\frac{17}{21}$   2) $4\frac{1}{24}$   3) $4\frac{3}{4}$   4) $3\frac{1}{10}$   5) $1\frac{21}{28}$   6) $3\frac{11}{18}$
7) 4   8) 2

**Exercises 42**
1) $(\frac{1}{5})$   2) $(\frac{5}{7})$   3) $(\frac{1}{2})$   4) $1\frac{1}{2}$   5) $1\frac{1}{2}$   6) $(\frac{1}{6})$
7) $(\frac{3}{4})$   8) $(\frac{7}{12})$

**Exercises 43**
1) $(\frac{5}{7})$   2) $(\frac{7}{15})$   3) $(\frac{1}{8})$   4) $1\frac{4}{5}$   5) $1\frac{3}{5}$   6) $1\frac{1}{2}$
7) $3\frac{4}{7}$   8) $(\frac{5}{7})$   9) $1\frac{1}{3}$   10) 1   11) $(\frac{1}{3})$   12) 0

**Exercises 44**
1) $1\frac{1}{3}$   2) $1\frac{8}{9}$   3) $6\frac{1}{8}$   4) $2\frac{5}{14}$   5) $(\frac{1}{6})$   6) $(\frac{5}{8})$
7) $2\frac{11}{15}$   8) $1\frac{5}{12}$

**Exercises 45**
1) $2\frac{11}{21}$   2) $(\frac{7}{12})$   3) $1\frac{19}{20}$   4) $1\frac{1}{18}$   5) $(\frac{5}{6})$   6) $(\frac{2}{15})$
7) $1\frac{17}{40}$   8) $(\frac{7}{30})$   9) $(\frac{51}{54})$   10) $2\frac{43}{56}$

**Exercises 46**
1) $1\frac{1}{3}$   2) $2\frac{3}{4}$   3) $1\frac{17}{35}$   4) $3\frac{1}{3}$   5) $(\frac{1}{2})$   6) $(\frac{1}{10})$
7) $(\frac{7}{8})$   8) $(\frac{9}{12})$

**Exercises 47**
1) $(\frac{2}{3})$   2) $(\frac{1}{3})$   3) 20   4) 12   5) 5   6) 3

## Exercises 48
1) $1\frac{4}{9}$  2) $4\frac{1}{6}$  3) $(\frac{3}{7})$  4) $(\frac{1}{5})$  5) $(\frac{1}{2})$  6) $(\frac{5}{12})$
7) $(\frac{4}{5})$  8) $3\frac{3}{5}$  9) $(\frac{11}{14})$  10) $16\frac{1}{2}$

## Exercises 49
1) $1\frac{2}{3}$  2) $2$  3) $7$  4) $(\frac{1}{3})$  5) $1$  6) $(\frac{3}{8})$

## Exercises 50
1) $4$  2) $7\frac{13}{16}$  3) $1\frac{5}{9}$  4) $(\frac{2}{27})$  5) $4$  6) $(\frac{3}{10})$
7) $1$  8) $4\frac{5}{7}$  9) $3\frac{3}{4}$  10) $4$

## Exercises 51
1) $4$  2) $(\frac{3}{8})$  3) $5\frac{1}{4}$  4) $4$  5) $(\frac{2}{3})$  6) $(\frac{3}{4})$

## Exercises 52
1) $(\frac{8}{9})$  2) $1\frac{1}{2}$  3) $2\frac{2}{15}$  4) $(\frac{16}{21})$  5) $1\frac{1}{3}$  6) $1\frac{1}{4}$
7) $1\frac{1}{7}$  8) $1\frac{8}{13}$  9) $12\frac{4}{7}$  10) $1\frac{1}{15}$

## Exercises 53
1) $(\frac{3}{4})$  2) $(\frac{1}{2})$  3) $7$  4) $(\frac{1}{7})$  5) $3$  6) $(\frac{2}{5})$

## Exercises 54
1) $22\frac{1}{2}$  2) $3\frac{1}{16}$  3) $(\frac{5}{12})$  4) $7$  5) $8\frac{4}{5}$  6) $8\frac{4}{9}$
7) $1\frac{7}{20}$  8) $2\frac{1}{4}$  9) $1\frac{1}{9}$  10) $1\frac{15}{49}$  11) $1\frac{9}{16}$  12) $1\frac{1}{32}$

## Exercises 55
1) $(\frac{12}{35})$  2) $1$  3) $1\frac{1}{2}$  4) $1\frac{1}{5}$  5) $2\frac{3}{4}$  6) $(\frac{5}{36})$


# CHAPTER 4

## Exercises 1
1) ¾, 75%  2) 1/4, 25%  3) 1/10, 10%  4) ½, 50%  5) 13/25, 52%  6) 63/100, 63%
7) 2/5, 40%  8) ¾, 75%

## Exercises 2
1) 2%  2) 15%  3) 60%  4) 380%  5) 275%  6) 5%
7) 120%  8) 28%  9) 20%  10) 150%

## Exercises 3
1) 3/100  2) 130/100  3) 19/100  4) 2518/100  5) 8/10000  6) 37/1000
7) 95/10000  8) 92/100  9) 2(1/2)  10) 9(99/1000)

# Answers

## Exercises 4
1) 27/29    2) 93.1%    3) 82%

## Exercises 5
1) 45%    2) 19%

## Exercises 6
1) 1%    2) 36%    3) 50%    4) 102%    5) 82%    6) 146%
7) 70%    8) 59.1%    9) 107.8%    10) 30.64%    11) 352.6%    12) 901%

## Exercises 7
1) 62.5%    2) 63%    3) 1.5%    4) 0.2%    5) 105%    6) 120%
7) 112.5%    8) 1220%

## Exercises 8
1) 0.02    2) 1.05    3) 0.47    4) 0.92    5) 0.32    6) 1.20

## Exercises 9
1) 9/100    2) 10(5/1000)    3) 12/100    4) 2(1/2)    5) 98/100    6) 1(5/10)

## Exercises 10
39.7%

## Exercises 11
1) 0.001, 1/1000    2) 0.0001, 1/10,000    3) 0.00101, 101/100,000    4) 0.0101, 101/10,000    5) 0.00293, 293/100,000
6) 0.0304, 304/10,000    7) 0.1081, 1081/10000    8) 0.0325, 325/10,000    9) 0.00305, 305/100,000    10) 0.00064, 64/100,000

## Exercises 12
1) 0.025    2) 0.1125    3) 27    4) 1.404    5) 0.001    6) 0.102
7) 0.6    8) 1.98    9) $38.40    10) 5.4    11) 0.6    12) 1.8

## Exercises 13
1) $0.06    2) 0.17    3) $72.00    4) 17.6    5) 0.02    6) 0.2
7) 12.75    8) 110    9) $22.75    10) $2.50    11) $6.97    12) 115
13) $1.20    14) $2.40

## Exercises 14
1) 0.5    2) 6    3) 22.08    4) 0.25    5) 120    6) 20
7) $9.00    8) 3.04    9) 11.6    10) 237.5

## Exercises 15
1) 114    2) $2.40    3) $0.12    4) $0.1125    5) $0.06    6) $0.72
7) $90.00    8) $1.44    9) 22    10) $0.098

## Exercises 16
1) 0.475lb    2) 12.5%

## Exercises 17
20%

Exercises 18
1)

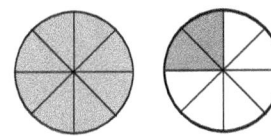

3)

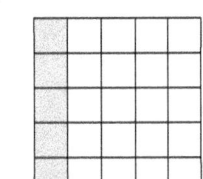

2) 81.25%

4) 48%

## CHAPTER 5

Exercises 1
1)   Ray      2)   Line segment   3)   Line

Exercises 2
1)   Similar   2)   Congruent

Exercises 3
1)   F      2)   D      3)   E      4)   C      5)   A      6)   B

Exercises 4
1)   Rotation   2)   Reflection   3)   Translation   4)   Rotation

Exercises 5
1)   B      2)   A      3)   C      4)   D      5)   E      6)   F

Exercises 6
1)   cone   2)   triangle   3)   sphere   4)   pentagon   5)   trapezoid   6)   circle

Exercises 7
1)   30 ft               2)   74 cm               3)   37 cm

Exercises 8
1)   $P = 20$ cm      2)   $P = 38$ m      3)   $P = 130$ in.      4)   $P = 36$ ft
     $A = 25$ cm$^2$        $A = 84$ m$^2$        $A = 720$ in$^2$        $A = 30$ ft$^2$

Exercises 9
1)   6.5 in      2)   5 cm      3)   6 in      4)   9 ft

Exercises 10
1)   G      2)   B      3)   A      4)   F      5)   C      6)   H
7)   D      8)   I      9)   E

Exercises 11
1)   F: 6, V: 8, E: 12      2)   F: 5, V: 5, E: 8      3)   F: 5, V: 6, E: 9

Exercises 12
1)   Triangular prism      2)   Square pyramid      3)   Cylinder

Exercises 13
1)   112 in.$^2$               2)   348 ft$^2$               3)   54 cm$^2$

### Exercises 14
1) 258 cm$^2$                    2) 166 yd$^2$                    3) 230 cm$^2$

### Exercises 15
1) B        2) D        3) A        4) F        5) C        6) E

### Exercises 16
1) B        2) A        3) D        4) C

## CHAPTER 6

### Exercises 1
1) 3, 14, 20, 12, 13        2) Cell phones        3) Desktop computers        4) 11

### Exercises 2
1) 4                    2) Soup                    3) Appetizers

### Exercises 3
1) Mode = 9            2) Mode = 70            3) Mode = 40
   Median = 7.5           Median = 80               Median = 40
4) Mode = no           5) Mode = no            6) Mode = 20
   Median = 152          Median = 13.5             Median = 20

### Exercises 4
1) Mean = 162.29       2) Mean = 3.7           3) Mean = 32
   Median = 163          Median = 4               Median = 30.5
4) Mean = 89.6         5) Mean = 7             6) Mean = 55
   Median = 90           Median = 6.5             Median = 55

### Exercises 5
1) 42                    2) It equals to 6 or less than 6        3) 111

### Exercises 6
1) 0-350        2) 50                    3) Main Salad        4) Main Salad
5) Fruit

### Exercises 7
1) 2, 6         2) 49                    3) 2            4) 51-100
5) 21

### Exercises 8
1) 0-21         2) 16                    3) 2            4) About 7
5) 32           6) 27                    7) R = 15, S = 1

Exercises 9

1)

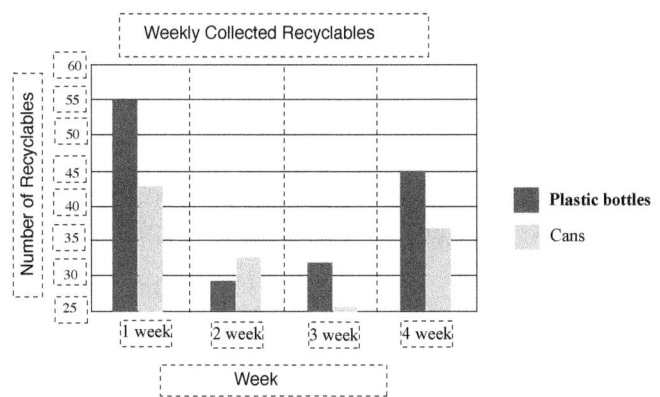

2)   1 week          3)   3 week          4)   It will be vary.

Exercises 10
1)   Memphis, TN    2)   Apr.            3)   Aug.            4)   Jan.
5)   Jan            6)   It will be vary.

Exercises 11
1)   Line graph     2)   Jan.            3)   May             4)   Jan, Feb., Sep.
5)   About $240     6)   $250            7)   It can be vary.

Exercises 12
1)   Jul.           2)   0-90            3)   10              4)   Jul.
5)   Mar. and Nov.  6)   50 or lower

Self-Test, Page 5
1)   C    2)   C    3)   A    4)   A    5)   A    6)   B
7)   C    8)   D    9)   C    10)  C    11)  C    12)  B
13)  C    14)  A

Self-Test, Page 11
1)   C    2)   A    3)   D    4)   B    5)   B    6)   B
7)   D    8)   A    9)   C    10)  D

Self-Test, Page 21
1)   D    2)   B    3)   A    4)   D    5)   C    6)   D
7)   D    8)   B    9)   A    10)  C    11)  A    12)  A
13)  A    14)  B    15)  B    16)  D    17)  A    18)  C

Self-Test, Page 36
1)   D    2)   B    3)   A    4)   B    5)   D    6)   A
7)   D    8)   C    9)   D    10)  A    11)  D    12)  B
13)  B    14)  C    15)  A    16)  D    17)  C    18)  C
20)  B    21)  A

## Self-Test, Page 47

| | | | | | | | | | | | |
|---|---|---|---|---|---|---|---|---|---|---|---|
| 1) | D | 2) | C | 3) | B | 4) | B | 5) | A | 6) | B |
| 7) | A | 8) | A | 9) | D | 10) | A | 11) | B | 12) | C |
| 13) | C | 14) | D | 15) | D | 16) | C | 17) | D | 18) | C |
| 19) | C | 20) | D | 21) | B | 22) | D | 23) | B | 24) | B |

## Self-Test, Page 54

| | | | | | | | | | | | |
|---|---|---|---|---|---|---|---|---|---|---|---|
| 1) | D | 2) | C | 3) | A | 4) | C | 5) | D | 6) | A |
| 7) | A | 8) | A | 9) | B | 10) | C | 11) | A | 12) | D |
| 13) | C | 14) | B | 15) | C | 16) | B | 17) | A | 18) | D |
| 19) | D | 20) | C | | | | | | | | |

## Self-Test, Page 63

| | | | | | | | | | | | |
|---|---|---|---|---|---|---|---|---|---|---|---|
| 1) | D | 2) | D | 3) | A | 4) | C | 5) | D | 6) | C |
| 7) | A | 8) | B | 9) | D | 10) | B | 11) | A | 12) | C |
| 13) | C | 14) | C | 15) | D | 16) | C | 17) | B | 18) | C |
| 19) | C | 20) | B | 21) | B | 22) | A | | | | |

## Self-Test, Page 71

| | | | | | | | | | | | |
|---|---|---|---|---|---|---|---|---|---|---|---|
| 1) | C | 2) | C | 3) | D | 4) | A | 5) | C | 6) | C |
| 7) | B | 8) | D | 9) | A | 10) | B | 11) | A | 12) | B |
| 13) | C | | | | | | | | | | |

## Self-Test, Page 76

| | | | | | | | | | | | |
|---|---|---|---|---|---|---|---|---|---|---|---|
| 1) | B | 2) | B | 3) | C | 4) | A | 5) | D | 6) | A |
| 7) | B | 8) | D | 9) | B | 10) | C | 11) | C | 12) | D |
| 13) | B | 14) | A | 15) | A | 16) | C | 17) | C | 18) | B |

## Self-Test, Page 85

| | | | | | | | | | | | |
|---|---|---|---|---|---|---|---|---|---|---|---|
| 1) | B | 2) | C | 3) | D | 4) | D | 5) | C | 6) | D |
| 7) | B | 8) | D | 9) | B | 10) | D | 11) | B | 12) | C |
| 13) | A | 14) | C | 15) | C | 16) | C | 17) | B | 18) | B |
| 19) | D | | | | | | | | | | |

## Self-Test, Page 93

| | | | | | | | | | | | |
|---|---|---|---|---|---|---|---|---|---|---|---|
| 1) | D | 2) | C | 3) | A | 4) | A | 5) | D | 6) | A |
| 7) | B | 8) | A | 9) | A | 10) | B | 11) | B | 12) | C |

## Self-Test, Page 102

| | | | | | | | | | | | |
|---|---|---|---|---|---|---|---|---|---|---|---|
| 1) | A | 2) | B | 3) | D | 4) | B | 5) | C | 6) | C |
| 7) | D | 8) | A | 9) | B | 10) | D | 11) | D | 12) | C |

## Self-Test, Page 109

| | | | | | | | | | | | |
|---|---|---|---|---|---|---|---|---|---|---|---|
| 1) | D | 2) | C | 3) | A | 4) | C | 5) | A | 6) | C |
| 7) | A | 8) | A | 9) | D | 10) | B | 11) | D | 12) | C |
| 13) | D | 14) | B | 15) | C | 16) | A | 17) | D | 18) | D |

## Self-Test, Page 121

| | | | | | | | | | | | |
|---|---|---|---|---|---|---|---|---|---|---|---|
| 1) | C | 2) | D | 3) | C | 4) | C | 5) | A | 6) | A |
| 7) | C | 8) | C | 9) | D | 10) | B | 11) | D | 12) | D |
| 13) | B | | | | | | | | | | |

## Self-Test, Page 130

| | | | | | | | | | | | |
|---|---|---|---|---|---|---|---|---|---|---|---|
| 1) | C | 2) | C | 3) | D | 4) | C | 5) | D | 6) | D |

| 7) C | 8) C | 9) A | 10) D | 11) B | 12) D |
| 13) B | 14) A | 15) D | | | |

**Self-Test, Page 137**

| 1) D | 2) A | 3) C | 4) C | 5) A | 6) B |
| 7) B | 8) A | 9) B | 10) C | 11) C | 12) D |

**Self-Test, Page 141**

| 1) A | 2) D | 3) B | 4) A | 5) B | 6) A |
| 7) C | 8) D | 9) A | 10) D | 11) C | 12) C |
| 13) A | | | | | |

**Self-Test, Page 148**

| 1) A | 2) A | 3) C | 4) A | 5) B | 6) A |
| 7) A | 8) A | 9) A | 10) B | 11) A | 12) A |

**Self-Test, Page 153**

| 1) B | 2) A | 3) A | 4) C | 5) C | 6) B |
| 7) D | 8) A | 9) D | 10) B | 11) B | |

**Self-Test, Page 157**

| 1) C | 2) D | 3) C | 4) B | 5) B |
| --- | --- | --- | --- | --- |

**Self-Test, Page 162**

| 1) A | 2) D | 3) C | 4) B | 5) D | 6) D |
| 7) D | 8) A | 9) B | 10) C | 11) A | 12) A |
| 13) D | 14) B | 15) D | 16) A | 17) B | 18) D |
| 19) D | 20) D | | | | |

**Self-Test, Page 167**

| 1) B | 2) A | 3) B | 4) D | 5) A | 6) C |
| 7) D | 8) D | 9) B | 10) C | 11) B | 12) A |

**Self-Test, Page 170**

| 1) A | 2) A | 3) B | 4) D | 5) C | 6) C |
| 7) B | 8) D | 9) D | 10) D | 11) A | 12) C |
| 13) D | 14) D | 15) B | 16) D | 17) C | |

**Self-Test, Page 177**

| 1) D | 2) A | 3) B | 4) C | 5) A | 6) B |
| 7) D | 8) B | 9) C | 10) D | 11) D | 12) D |

**Self-Test, Page 182**

| 1) C | 2) C | 3) D | 4) A | 5) B | 6) A |
| 7) C | 8) B | 9) B | 10) D | 11) A | 12) D |
| 13) C | | | | | |

**Self-Test, Page 188**

| 1) C | 2) D | 3) A | 4) C | 5) B | 6) C |
| 7) D | 8) C | 9) B | 10) D | 11) D | 12) B |
| 13) D | 14) C | | | | |

# Visit us at WWW.IQMATHS.com

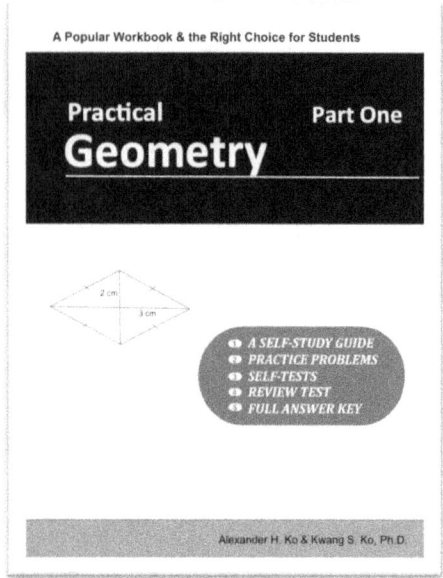

ISBN: 978-1-5232673-6-1

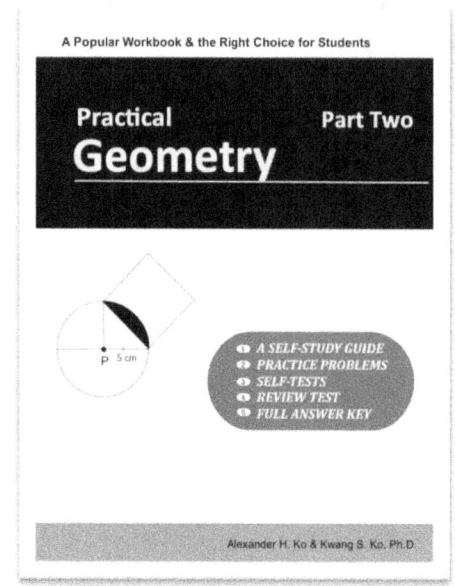

ISBN: 978-1-5233620-1-1

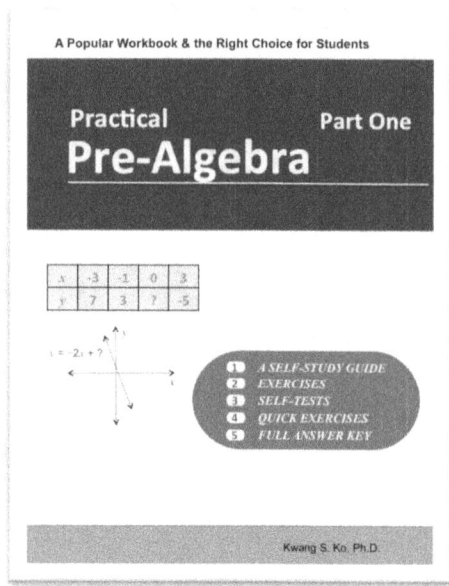

ISBN: 978-1-5233628-6-8

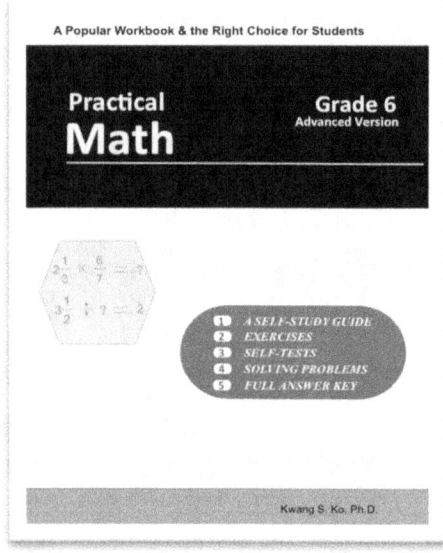

ISBN: 978-1-5233628-9-9

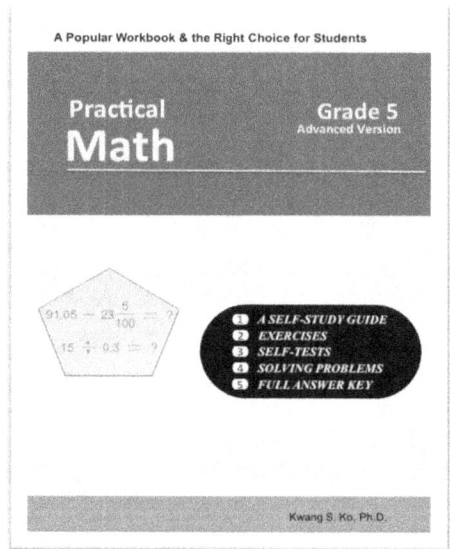

ISBN: 978-1-5233630-1-8

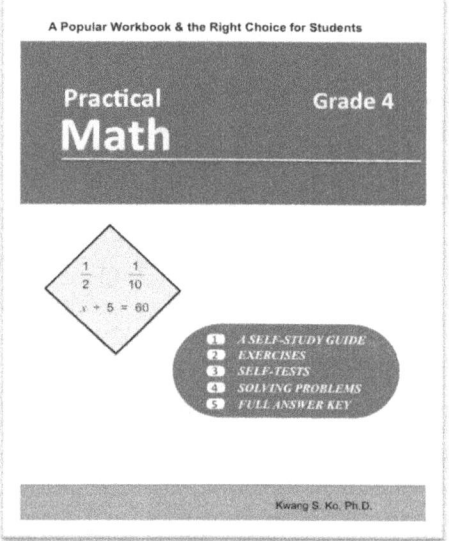

ISBN: 978-1-5233630-2-5

Other books are sold at WWW.IQMATHS.com.

www.ingramcontent.com/pod-product-compliance
Lightning Source LLC
Chambersburg PA
CBHW080008210526
45170CB00015B/1925